PRAISE FOR ROBERT OSSERMAN'S
Poetry of the Universe

"Wow. *Poetry of the Universe* manages to pack a lot in . . . a masterful job."

—*Los Angeles Times*

"Though the light of science and the light of art are inseparable and the same, their bearers speak different languages and only the best among them understand that they are engaged in the same enterprise. In *Poetry of the Universe,* Robert Osserman enriches the tradition of unifying the two disciplines by speaking his own language with extraordinary clarity and accessibility, by burning off the clouds."

—MARK HELPRIN

"Thoroughly delightful. It shows how mathematics and our understanding of the universe evolve together. Osserman's lucid explanations and passion make this book daring in scope and yet richly personal."

—GEORGE F. SMOOT, astrophysicist,
University of California at Berkeley,
author of *Wrinkles in Time*

"An elegant introduction to many of the beauties of mathematics and their relationship to the physical world."

—ROGER PENROSE, author of *The Emperor's New Mind*
and Rouse Ball Professor of Mathematics, Oxford University

"Elegant . . . the atmosphere is leisurely and there is much more emphasis on the integration of the history of ideas in mathematics with the surrounding culture than on simplified expositions."

—*Nature*

"*Poetry of the Universe* sets forth with admirable clarity the history of the mathematical conquest of the world around us."

—DIANE MIDDLEBROOK, author of *Anne Sexton*

"A celebration of the genius of man at every level . . . it should be required [reading] by anyone who wishes to have a greater understanding of mankind, our genius and our possibilities."

—*South Bend Tribune*

"[*Poetry of the Universe*] will be appreciated by all who read it."

—*Capital Times*

"A wonderful book. It shows the extent to which the universe becomes understandable through the medium of mathematics."

—PHILIP J. DAVIS, co-author of the American Book Award–winning *The Mathematical Experience*

"Essential reading for those educated in the liberal and fine arts who have never had the opportunity to appreciate the beauty of mathematics and physics."

—*Library Journal*

Poetry

of the

Universe

A Mathematical Exploration of the Cosmos

Poetry

of the

Universe

Robert Osserman

Anchor Books
Doubleday
New York London Toronto Sydney Auckland

AN ANCHOR BOOK
PUBLISHED BY DOUBLEDAY
a division of Bantam Doubleday Dell Publishing Group, Inc.
1540 Broadway, New York, New York 10036

ANCHOR BOOKS, DOUBLEDAY, *and the portrayal of an anchor are trademarks of Doubleday, a division of Bantam Doubleday Dell Publishing Group, Inc.*

Book design by Terry Karydes

Poetry of the Universe *was originally published in hardcover by Anchor Books in 1995.*

The Library of Congress has cataloged the Anchor hardcover edition of this work as follows:
Osserman, Robert.
Poetry of the universe : a mathematical exploration of the cosmos / Robert Osserman.
p. cm.
Includes bibliographical references and index.
1. Mathematics—Popular works. I. Title.
QA93.O87 1995
530.1′5—dc20 94-27971
CIP

ISBN 0-385-47429-6

PRINTED IN THE UNITED STATES OF AMERICA
FIRST ANCHOR BOOKS TRADE PAPERBACK EDITION: FEBRUARY 1996

1 3 5 7 9 10 8 6 4 2

Dedication

To my wife, Janet Adelman, for moral, spiritual, nutritional, and syntactical sustenance during the long gestation of this book, and to my sons, Brian and Stephen, for their active interest, assistance, and understanding.

Contents

EUCLID ALONE HAS LOOKED ON BEAUTY BARE.

—Edna St. Vincent Millay

PURE MATHEMATICS IS, IN ITS WAY,
THE POETRY OF LOGICAL IDEAS.

—Albert Einstein

WE HAVE HEARD MUCH ABOUT THE POETRY OF
MATHEMATICS, BUT VERY LITTLE OF IT HAS YET BEEN SUNG
. . . THE MOST DISTINCT AND BEAUTIFUL STATEMENT OF
ANY TRUTH MUST TAKE AT LAST THE MATHEMATICAL FORM.

—Henry David Thoreau

PREFACE

On April 24, 1992, newspapers around the world reported an event that was hailed as "one of the major discoveries of the century"—what some would call "the missing link" and "the Holy Grail" of cosmology. The discovery was presented in the form of a picture that was in essence a snapshot of the universe at a dramatic moment in its evolution—the moment that space began. Before the time of the picture, there was only a conglomeration of elementary particles in a state of continual creation and annihilation. Then electrons and protons combined to form atoms of matter. For the first time there was space between the atoms, allowing light and other forms of radiation to travel freely. The "snapshot" depicts the pattern of rays that have reached us after traveling through space from that moment to the present. What was electrifying to scientists who had been studying those rays—the so-called cosmic microwave background radiation—was that there *was* a pattern in the picture. After decades of frustration at trying to detect even a ripple of variation in the apparently featureless sea of uniform background radiation, they were now successful in finding a possible link between the undifferentiated primeval "soup" predicted by the big bang theory of the creation of the universe and the later evolution into the highly differentiated stars and galaxies of the universe as we know it today. But reporters attempting to explain the precise nature of the picture were faced with at least one insurmountable obstacle: neither they nor their readers were prepared for the paradoxical nature of an image that depicts simultaneously a view outward in all directions from the earth and inward in all directions toward the big bang.

The big discovery of 1992 recalls in both obvious and more subtle ways the "discovery" of America just five hundred years earlier. If we look further back to the year 1000 A.D., Europeans generally pictured the earth as flat. It took great efforts of the imagination over the subsequent centuries to grapple with and understand the implications of a spherical earth, and to appreciate why people at the opposite side who were hanging upside down by their heels did not either fall off or suffer from perpetual headaches. The voyages of Columbus and those following in his wake gave a sense of reality and concreteness to the theoretical view of a round earth that had gradually been established before his time.

Now, as we approach the year 2000, there are few people left who believe in a flat earth, but most of the world's population continues to think in terms of a flat universe. Just as everyday experience led us to think of the earth as flat or planar rather than curved, so does our perception of the world around us lead us to view space as flat or "euclidean." It requires as great an effort of the imagination in the twentieth century to conceive of curved space as it did a thousand years ago to conceive of

The 1992 picture of variation in the cosmic background radiation (courtesy of George Smoot, NASA, and the COBE satellite).

the earth as a gigantic ball somehow suspended or floating freely in an even more gigantic expanse of space. Nevertheless, the evidence is overwhelming that space is indeed curved, and only in that context can the cosmic microwave photograph of 1992 be fully understood.

What is the shape of the universe, and what do we mean by the curvature of space? One aim of this book is to make absolutely clear and understandable both the meaning of those questions and the answers to them. Little or no mathematical background is needed; the book takes the reader from the easily understandable mathematical ways of evaluating the world to those notions that are unfamiliar and further removed from everyday experience, while conveying the excitement and the power of the mathematical ideas that form the core of modern cosmology. The history and evolution of these ideas are often as fascinating as the ideas themselves, and are presented in an unfolding chronological narrative along with glimpses of the lives and personalities of some key players involved in the story. For those who wish to know more about the technical and mathematical underpinnings of the notions introduced, a section of notes at the back of the book provides further details as well as references to sources for further reading.

Prelude

Imagine sailing on a clear but windy day. The surface of the water is choppy and bright blue, reflecting and intensifying the color of the sky. Suddenly the weather changes—the wind ceases, the skies cloud over, and the surface of the ocean becomes calm and smooth. The water itself turns green and transparent, allowing a glimpse of a coral reef and a whole new world of colorful activity down below. If you dive below the surface to get a closer look, you'll find that your 20/20 vision in the air above produces only a blur underwater. But if someone provides you with a pair of goggles, then suddenly the world beneath the surface becomes just as clear and even more beautiful than your original view of the surface from above.

Imagine, similarly, going out to the middle of a desert on a clear and moonless night, far from city lights. Against a pure black background, the stars, the planets, the nebulae, the constellations, the Milky Way, stand out in a dizzying array. With the aid of a telescope, more and more exotic sights appear: majestic spiral galaxies, great spherical balls of light and color from past supernova explosions. Out of the "cosmic static" from the first radio telescopes, more and more refined astronomical instruments bring images of pulsars and quasars, as well as the ubiquitous cosmic microwave background radiation. These are all surface appearances, however, on the magnificent ocean of the cosmos. What lies below the surface and beyond one's field of view, what provides the underlying structure out of which all these phenomena evolved, cannot be seen without the necessary equipment: a pair of mind-goggles allowing the imagination to function outside its natural boundaries.

The goal of this book is to provide those mind-goggles, enabling the reader to move freely in the unfamiliar world of curved space-time. One cannot expect to plunge all at once into that world, but a bit of patience and persistence and the right tools will bring substantial rewards as new and unsuspected vistas emerge. Beyond that, the book is a celebration of the human imagination—the facility to make the kind of mental leaps without which the impact of the outer world on our senses would be mostly noise. Mathematical imagination and imagery, closely linked, provide the vision that allows us to see the hidden but exquisite structure below the surface.

Chapter 1

Measuring the Unmeasurable

Thy shadow, Earth, from Pole to Central Sea,
Now steals along upon the Moon's meek shine
In even monochrome and curving line
Of imperturbable serenity.

—Thomas Hardy,
"At a Lunar Eclipse"

Over two thousand years ago, the philosopher-scientists of ancient Greece embarked on a project that was as daunting for those days as exploring the boundaries of the solar system would be today. It was to determine the size and shape of the entire earth. To the ancient Greeks, the earth was unimaginably big. Neither the Greeks nor any of the civilizations they came into contact with had roamed or sailed over more than a fraction of it. To go from the minuscule portions of the earth that could be directly measured to the immensities of unexplored and even undreamed-of distant lands required great feats of ingenuity. It also required the systematic development of an entirely new branch of learning that the Greeks would call *geometry,* meaning literally: "measuring the earth."

One of the best-known names from the early history

of geometry is that of Pythagoras, whose lifetime spanned most of the sixth century B.C. But long before Pythagoras, the Egyptians had devised a simple method for constructing perpendicular lines, such as those framing the base of a pyramid. They used knots placed at equal intervals along a rope to divide the rope into lengths of 3, 4, and 5 spaces between knots. Placing pegs in the ground so that the rope formed a triangle, when stretched taut around the pegs, with side lengths of 3, 4, 5 respectively, they found that the angle between the sides of length 3 and 4 was a right angle, or 90 degrees. They also found that different side lengths would serve the same purpose when a certain condition was met. The key to getting a right angle was to have the square of the longest side equal to the sum of the squares of the other two sides, a relationship that we know as the "Pythagorean theorem." The Babylonians were also aware of this relationship. In fact, over a thousand years before Pythagoras, at the time of Hammurabi, "the Lawgiver," the Babylonians had developed mathematics to a much higher level than the Egyptians, including a more sophisticated system for representing numbers and some basic algebra, as well as geometry. Not only did they seem to be aware of the "Pythagorean theorem" but they also provided a long list of number triples, including such unlikely ones as (65, 72, 97) and (119, 120, 169), all representing the sides of a right triangle.

Why then was the theorem named after the latecomer Pythagoras? Despite the priority of the Egyptians and Babylonians, they gave no indication of having thought of the key mathematical notion of a proof.

Pythagoras's name became attached to the theorem because he was reputedly the first person to provide such a proof. As it happens, there is no direct evidence that he did so. (It is not known if Pythagoras left anything in written form; if he did, none of his writings have survived to the present.) It seems most probable that the first proof of the "Pythagorean theorem" originated with his followers, the "Pythagoreans," sometime in the following century.

Euclid, who became the most famous of all Greek mathematicians, was born over two hundred years after Pythagoras. During the period between Pythagoras and Euclid, geometry developed on two parallel tracks. One consisted of the detailed study of particular shapes, such as triangles, rectangles, and figures bounded by circular arcs. The other was the development of the method of proof and the process of deductive reasoning, leading to new discoveries that would not have been found by direct observation. By the time Euclid arrived on the scene, a sizable body of geometric lore had accumulated.

The details of Euclid's life, even more than of Pythagoras's, remain shrouded in obscurity. Virtually all that can be stated with certainty is that he lived and worked in Alexandria during a period around 300 B.C. Unlike Pythagoras, however, he left writings that not only have survived to the present but have become the basis of much of modern science, as well as a model for all of mathematics.

The monumental work for which Euclid is best known is *The Elements,* a mathematical compendium in thirteen books, of which five are devoted to the geometry of two-dimensional figures, three are devoted to the geometry of three dimensions, and the remainder to other subjects.

Euclid's *Elements* made a deep impact on the psyche of the Western world. Originally viewed as both a tool and a model for research in mathematics and other sciences, *The Elements* gradually evolved into a basic component of a standard education—a piece of intellectual equipment that every young student was expected to wrestle with and internalize. The fascination of *The Elements* has at least four distinct components. First, there is the sense of certainty—that in a world full of irrational beliefs and shaky speculations, the statements found in *The Elements* were proven true beyond a shadow of a doubt. And although certain features in both the assumptions and the methods of reasoning used by Euclid have been questioned over the centuries, the astonishing fact is that after two thousand years, nobody has ever found an actual "mistake" in *The Elements*—that is to say, a statement that did not follow logically from the given assumptions. The second feature is the power of the method. Starting from a very few explicitly laid-out assumptions, Euclid produced a dazzling series of consequences. Third is the display of ingenuity employed in the proofs—not so different from the sort of ingenuity that adds to the appeal of a well-crafted detective story. Finally, the objects of the reasoning in the first books of *The Elements* are geometric shapes that have an aesthetic appeal of their own, quite apart from any formal reasoning that may be applied to them. Some combination of these features led Edna St. Vincent Millay to the sentiments expressed in her poem: "Euclid alone has looked on Beauty bare."

A geometrical figure.

Of all the shapes that were studied by mathematicians, one held a special fascination: the circle. Like euclidean geometry as a whole, this one shape—the circle—was destined to play a powerful role—for better and for worse—in all future attempts to describe the shape and the workings of the world and the universe.

How did the concept of a circle first enter human awareness? There are surprisingly few places in nature where one sees a true circle. The most striking example is undoubtedly the sun, a daily presence, even if too bright to view directly except when near the horizon or when filtered through a thin layer of clouds or fog. In some ways even more awe-inspiring is the full moon, taking shape gradually and transforming itself into a perfect circle once every twenty-eight days. Another, indirect example that star watchers became aware of is the course of the stars overhead each night, describing circular arcs across the sky, a pattern that is most noticeable for stars in the vicinity of the North Star. One of the occurrences of a circle on earth is the beautiful pattern of circular ripples caused by the first few raindrops on a still pool of water, or a pebble tossed into a quiet pond. For someone standing at the edge of the sea, or at the stern of a boat, the horizon itself takes the shape of an immense circle.

Perhaps it was the circular shape of the horizon that provided the first clue to the shape of the earth. The first concrete evidence of its shape in early times was not the result of direct observations of the earth itself, but rather came from watching the night skies. Although it is impossible to say when the first stargazers grasped the significance of their two key observations, both were noted by Aristotle, in the fourth century B.C.

The first observation had to do with lunar eclipses, the result of the sun, the earth, and the moon all lining

Four stages in a lunar eclipse, starting with the near-perfect circle of the full moon, and then successively revealing larger and larger circular bites out of the moon as the earth's shadow gradually moves across the face of the moon.

up so that the earth temporarily blocks the sun's light from reaching the moon. The shadow of the earth gradually moves across the face of the moon, and it is clearly circular.

The second piece of evidence was more roundabout, but even more convincing. It required observing the skies not from a fixed point on earth, but from a number of places of differing latitudes. What became apparent was that as one travels to the south, the familiar constellations in the north gradually appear lower in the sky, while those in the south appear higher. Furthermore, new constellations, never seen at higher latitudes, appear near the horizon. The farther south one travels, the higher these new constellations appear in the sky, and the greater the num-

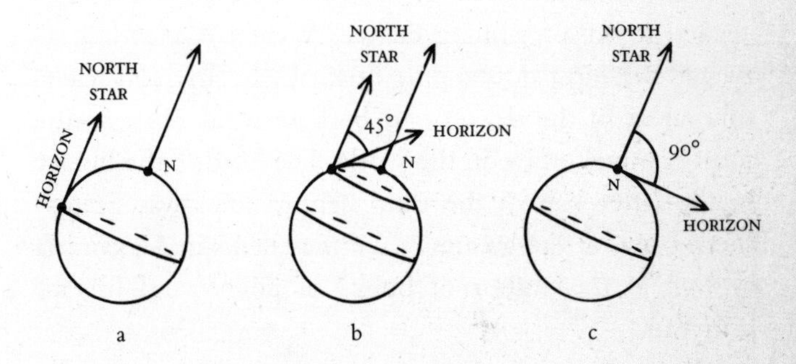

a. For an observer on the equator, the North Star would lie directly on the horizon.
b. At latitude 45°, an observer would see the North Star 45° above the horizon, or halfway from the horizon to the point directly overhead.
c. At the North Pole, the North Star would appear directly overhead.
The North Star is never visible from any location south of the equator.
(Note that all these statements are only approximately true. They would be exact if the North Star were precisely overhead at the North Pole, but it is actually off center by about 1°.)

ber of new constellations that come into view. It eventually became apparent that such changes were exactly what would be expected if the earth were spherical. And so, over two thousand years ago, the idea of a flat earth simply did not fit the observed facts, and had to be discarded. The more challenging question was not the qualitative one of the shape of the earth, but determining its size. How could one measure the whole earth, when the immense expanses of the oceans formed impenetrable barriers to travel? A most ingenious answer was provided by Eratosthenes of Alexandria.

Alexandria was founded at the delta of the Nile in northern Egypt, where the river empties into the Mediterranean Sea, by Alexander the Great, who wanted a city to match the grandeur of his own ambitions. He succeeded to an astonishing degree. Ancient Alexandria attracted the most outstanding literary, scholarly, and scientific talent of the day, in part because of its library—the most comprehensive in the world. The head of the library in the latter half of the third century B.C. was Eratosthenes, one of the greatest scientific talents in Alexandria as well as the author of books of poetry and literary criticism.

Eratosthenes' method for determining the size of the earth rested on three elements. The first was a bit of elementary geometry which will be explained in a moment. The second involved a serendipitous geographic fact regarding a city on the Nile River in southern Egypt called Syene in those days, now known as Aswan. The third was an absurdly simple apparatus called a *gnomon.*

The gnomon had been in use for a very long time. It consisted of a vertical stick placed on a level piece of ground. The gnomon was a device that allowed one to follow the sun's shadow as the sun moves across the sky. Although the gnomon cannot be used to tell time in the manner of its more advanced cousin, the sundial, it does provide a surprising amount of useful information.

First, the gnomon gives the exact time once a day, at the moment that the sun is highest in the sky and the shadow of the gnomon is the shortest—at noon. In addition, it acts as a compass, since the shadow at noon points due north (at least for Europe and most of the Northern Hemisphere, including Alexandria. In any event, regardless of which hemisphere one is in, the sun's shadow at noon always lies on the north-south axis.)

The gnomon also serves as a primitive calendar, determining two key days of each year: the summer and winter solstices. If one places a mark where the shadow ends at noon on each day of the year, one finds that in winter, when the sun is low in the sky, the shadows are longer, while in the summer, with the sun high in the sky, the shadows are shorter. The shadow at noon goes through a yearlong cycle, from the shortest noon shadow in summer, gradually reaching its greatest length six months later, and then shortening again over the succeeding six months. The day on which the noon shadow is shortest, and the sun is highest, is called the *summer solstice.* The day six months later when the sun is lowest and the noon shadow the longest is known as the *winter solstice.* Counting the number of days from solstice to solstice also provided one of the earliest accurate measurements of the length of the year.

Finally, the gnomon could be used to determine the altitude of the sun—that is, the angular distance of the sun above the horizon at any given moment (at least on sunny days). All one had to do was measure the length of the shadow and the length of the stick. By drawing a right triangle to scale with those measurements, one can measure the angle opposite the shadow, and that angle will indicate how far off the sun's direction is from an overhead, vertical direction.

These uses of the gnomon were well known to Eratosthenes and his contemporaries. But it was the for-

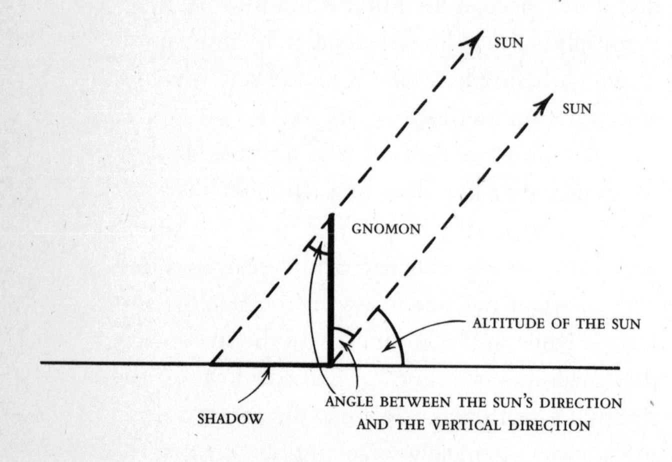

Gnomon and shadow.

tuitous geographical properties of Aswan that gave Eratosthenes his inspiration for determining the size of the earth. Aswan is almost due south of Alexandria. It also enjoys the special privilege of having the sun pass directly overhead at one moment of each year: at noon on the summer solstice. At that one moment each year, a gnomon in Aswan casts no shadow at all. (Aswan lies almost exactly on the Tropic of Cancer, the name given to the circle—about 23½ degrees above the equator—where the sun passes directly overhead at noon on the summer solstice.)

By combining these facts with some simple but clever geometric reasoning, Eratosthenes was able to produce his

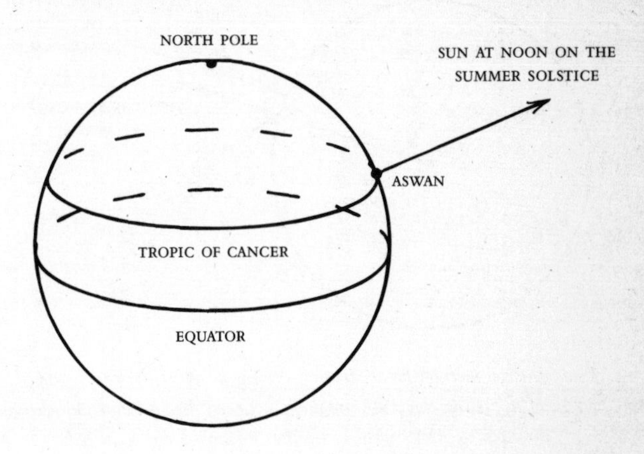

Tropic of Cancer is the name of a circle of latitude about 23.5 degrees above the equator.

remarkable pièce de résistance: the circumference of the earth. At noon on the summer solstice, he simply used his gnomon to determine the angle between the sun and the vertical direction in Alexandria. Since the sun at that moment is directly overhead at Aswan, he thereby knew the angle between the vertical directions at Alexandria and at Aswan. He found that angle to be ¹⁄₅₀ of the circumference of a circle. That meant that the entire circumference of the earth is 50 times the distance between Alexandria and Aswan. Since the distance from Aswan to Alexandria is roughly 500 miles, by today's measurements, the earth must be approximately 25,000 miles around.

The brilliant simplicity of Eratosthenes' method is not diminished by the fact that his estimate involves several

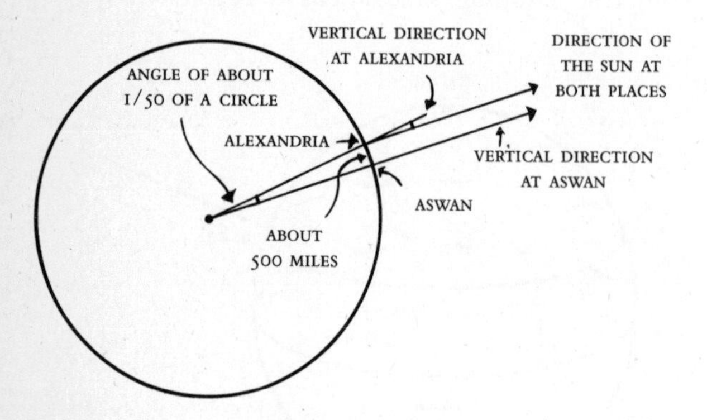

Eratosthenes' method for measuring the earth: when the sun is directly overhead at Aswan, measure the angle between the sun and the vertical direction at Alexandria using the shadow of a vertical pole.

inaccuracies and uncertainties: first, measuring the angle between the direction of the sun and the vertical direction could be done only approximately; second, Aswan is not exactly due south of Alexandria, but only roughly so; third, it would have been difficult or impossible to obtain an accurate measure of the distance between the two cities; and finally, there is considerable uncertainty about how to interpret ancient units of measurements in modern terms. Large distances were given in terms of *stades*—the length of a stadium. According to Eratosthenes, the circumference of the earth was 250,000 stades. The length of a "stade" was standardized at 600 "feet," but the length of a foot was not standard, and varied by 10 percent or more. The figure of 25,000 miles for the earth's circumference results from choosing a value at the low end of the scale for the length of a stade. The net effect was that Eratosthenes' calculation might on several counts be termed a "ballpark estimate" rather than a scientifically precise measurement. Nevertheless, it provides dramatic testimony to the ability of simple but ingenious geometric reasoning to succeed where a direct approach—involving traversals of two polar regions and an ocean—was well beyond the realm of possibility.

Eratosthenes' estimate of the size of the earth was the most famous but by no means the first such estimate. In the previous century, Aristotle cites a figure that he attributes to unnamed mathematicians, possibly from an even earlier epoch. These and subsequent estimates of the distance around the globe were to play an important role centuries later during the age of exploration. Furthermore, the kind of reasoning used by Eratosthenes was to play an even larger role in the long-term efforts to understand the shape and scope of the entire universe.

The ancient Greeks' attempts to measure the circumference of the earth naturally led to related questions, such as how to determine the size of the earth's diameter. If direct measurement of the distance along the surface seemed impossible, then measurements straight through the center of the earth were in the realm of pure fantasy. Again the answer was provided by geometry.

One of the basic properties of circles is that they all look alike; they can be large or small, scaled up or down in size, but since all parts scale alike, ratios such as that of the circumference to the diameter remain the same for all circles. The only question is: What is that ratio? It was known early on (and mentioned in the Bible) that the answer is approximately 3. (More accurate approximations, such as $3\frac{1}{8}$, were known much earlier to the Babylonians and Egyptians.) In modern times the Greek letter π, or pi, is used to denote that ratio, since π is the first letter in Greek of "perimeter" (meaning "to measure around"). The first careful calculations of the value of π were made by the greatest scientist of antiquity—Archimedes, a contemporary of Eratosthenes. He showed that the value of π was somewhere between $3\frac{10}{71}$ and $3\frac{1}{7}$. That means that if the distance around the world, starting and ending at Alexandria, is 25,000 miles, then the distance straight through is somewhere between 7,955 and 7,960 miles—a remarkably small margin of error.

Thus, over two thousand years ago the size and shape of the earth were pretty well established. Unfortunately,

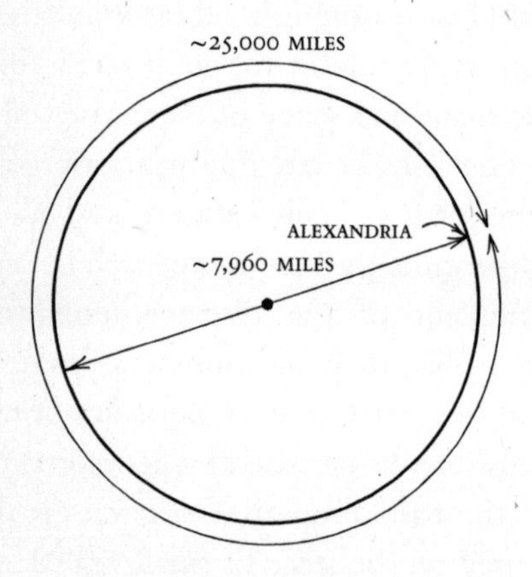

"Measuring" the diameter of the earth.

with the crumbling of the ancient civilizations, a thousand years of accumulated wisdom was lost to the European continent. By good fortune, the decline of the West coincided roughly with the rise of Arabic civilization and culture, and much of ancient knowledge was translated and transferred there. Along with it came the desire to refine what was known. One example was the astonishing feat of al-Kashi of Samarkand, who in 1424 carried Archimedes' method of computing π to undreamed-of lengths, determining its value to sixteen decimal places. He did so not only for the sheer joy of calculating far beyond what anyone had calculated before but also for a very specific purpose: to determine the circumference of the universe to within the width of a horsehair. To get a sense of the degree of accuracy he achieved in computing the value of π, if the earth were a perfect sphere with a circumference of 25,000 miles, then al-Kashi's estimate for π would determine the diameter of the earth to within less than one ten-millionth of an inch. The only distantly comparable achievement up to that time was the value $355/113$ or $3\,16/113$ discovered in the fifth century by the Chinese mathematician Tsu Ch'ung-chih, which gives the correct value of π to six decimal places.

Samarkand, as it happens, was one of the great centers of civilization during the time that al-Kashi lived and worked there. The Mongol conqueror Tamerlane made it his capital city. His grandson Ulug-Beg, who eventually succeeded him, prided himself on his prowess in mathematics and astronomy. He built an observatory in Samarkand where al-Kashi worked, and where the greatest catalog of stars to that time was compiled. Many of our modern star names are of Arabic origin.

Samarkand is situated in the southern part of modern Uzbekistan, a region that produced a remarkable percentage of the world's mathematical talent during the period after the decline of Greek civilization and before the modern period. Two major figures came from an area just south of the Aral Sea, known in those days as Khwarizm. One was the brilliant al-Biruni, who was born in 973 and lived well into the next century. We shall encounter his name later on. The other, even more renowned, was known simply as al-Khwarizmi, after his place of origin. His name metamorphosed into the modern word "algorithm" and the title of one of his books gave us the word "algebra." He also wrote the first treatise in Arabic on "Hindu arithmetic," which in turn was the first to be translated into Latin and was part of the thrust that led to the universal adoption of "Arabic numerals" in the West.

During the reign of the caliph al-Mamun, from 813 to 833, al-Khwarizmi worked in a kind of research library-institute founded by al-Mamun, called the House of Wisdom. One of al-Mamun's projects in which al-Khwarizmi participated was a careful estimate of the earth's circumference by means of a direct measurement of one degree of latitude on the earth's surface. A team of surveyors went to a large flat plain about 200 miles north of Baghdad (not far from the biblical town of Nineveh). They then traveled due north to a point where the height of the sun at midday was exactly one degree less than when they started. They found the distance traveled to be about 57 miles. Since the number of degrees in a full circle is 360, they concluded that the circumference of the earth must be 360 times the distance covered by the survey team, or some 20,500 miles. The unit of a "mile" then in use was somewhat longer than today's "mile," so that this estimate of the earth's circumference loses a bit in translation. But that matters little. What is important is that in the early ninth century, the spherical shape of the earth was accepted as simple fact in the world of Islamic science. The curvature of the earth was known to account for the varying altitudes of the stars and the sun as one traveled north or south, and by measuring the change in the angle of the stars or the sun above the horizon over a precisely measured distance on the earth, ninth-century scholars were able to determine the size as well as the shape of the earth.

But most Europeans around the year 1000 were oblivious to all the discoveries in the Near and Far East, or, for that matter, to the accumulated knowledge of the ancient Greeks more than a millennium before; they were unable to see beyond their horizons either literally or figuratively, and for them the world was flat, the universe impenetrable.

Chapter II

Encompassing the Earth

What god soever this division wrought,
And every part to due proportion brought;
First, lest the earth unequal should appear,
He turned it round, in figure of a sphere.

—Ovid, *Metamorphoses,* first decade A.D.
Translation by George Sandys, 1626

One of the enduring myths of the Western world is that in order to gain support for his expeditions, Christopher Columbus had to first overcome a pervasive belief that the earth was flat rather than round and that by attempting to sail west to Asia he would risk sailing off the edge of the earth. The myth undoubtedly stems in part from a compression of the historical past, conflating the early Middle Ages, when a belief in a flat earth was indeed widespread in Europe, with the late Middle Ages—centuries later—by which time Europe had caught up with, and partially surpassed, the state of knowledge of ancient Greece and medieval Islam.

Ptolemy—astronomer, geographer, and mathematician—lived in Alexandria over a thousand years before Columbus, during the height of the Roman Empire. During the second century A.D., he worked to consolidate and extend the scientific accomplishments of the previous centuries. Among his achievements was the completion of two works that were considered by Islamic scholars and Renaissance Europeans to provide the definitive descriptions of the heavens and the earth. The first book came to be known by the Latinized version of its part Arabic, part Greek title, *Almagest,* meaning "the greatest." The *Almagest* became for astronomy what Euclid's *Elements* was for geometry: the definitive treatise on the subject for over a thousand years.

In the same way, Ptolemy's *Geography* became the standard reference on that subject, to which all others deferred. Although one would expect the subjects of geometry, "measuring the earth," and geography, "describing the earth," to be indissolubly linked, their early histories were quite distinct. Geographical information was both scanty and unreliable, and maps were not meant to be interpreted too literally. Eratosthenes is credited with being one of the first to introduce mathematical methods into mapmaking. His methods were refined by Hipparchus, one of the great astronomers of antiquity, and also the first, to our knowledge, to develop trigonometry: the systematic study of relations between angles and side lengths in a triangle. Ptolemy made use of previous methods and discoveries, fully utilizing geometry in the service of geography. He wrote:

> In Geography one must contemplate the extent of the entire earth, as well as its shape, and its position under the heavens, in order that one

> may rightly state what are the peculiarities and proportions of the part with which one is dealing . . .
>
> It is the great and exquisite accomplishment of mathematics to show all these things to human intelligence . . .

After the fall of Rome, mapmaking in Europe reverted to its earlier, more fanciful state, based on beliefs and hearsay, rather than on facts and science. It was not until the thirteenth century that Ptolemy's *Geography* once more became available, but at that time only in the original Greek, a language that was not widely known. Another two hundred years passed before it was translated into Latin; the first printed version dates from 1472. Columbus himself owned a copy printed in 1479.

The spherical shape of the earth is accepted as established fact in Ptolemy's *Geography.* Several subsequent books written between 1200 and 1500 continued the discussion of the shape of the earth; the title alone of one of the most notable of them provides clear evidence of how educated people in the fifteenth century viewed the world. It was called simply *The Sphere.* The book's author was known as Sacrobosco, a Latinized version of the name of an Englishman, John of Holywood, who wrote a number of books in the early part of the thirteenth century. Probably no other textbook in history has been as successful as Sacrobosco's—it remained in print and was still in use five hundred years after it was written. (If one were to count Euclid's *Elements* as a textbook it would undoubtedly own the record for longevity, but it was written in a quite different spirit and certainly was not intended as a text.)

What Sacrobosco did in his book was to adapt key passages from the *Almagest* of Ptolemy, add some newer material, and omit a number of technicalities, to arrive at a more accessible description of the workings of the universe, as best understood at the time. As the title indicated, the "sphere" was the key to everything. The earth was a sphere lying inside the great sphere of the fixed stars, while the sun, moon, and planets were attached to intermediate spheres. The evidence that the earth was round and not flat was taken straight from Ptolemy:

> If the earth were flat from east to west, the stars would rise as soon for westerners as for orientals, which is false. Also, if the earth were flat from north to south and vice versa, the stars which were always visible to anyone would continue to be so wherever he went, which is false. But it seems flat to human sight because it is so extensive.

The Sphere includes a variant of Eratosthenes' estimate of the circumference of the earth and a calculation of the earth's diameter, using the value $22/7$ for π.

How influential was Sacrobosco's book? One of the requirements to attain the degree of the Licentiate in Paris in 1366 was interpreted by the faculty to mean that one must attend a series of lectures on *The Sphere* and one other book. *The Sphere* was one of the requirements for the A.B. degree in Vienna in 1389, as it was at Oxford in

1409 and at Erfurt in Germany in 1422. At least two other major universities at the time, Prague and Bologna, included *The Sphere* as required reading in their curriculum.

By Columbus' day the view that the earth was spherical was clearly neither idiosyncratic nor controversial. Ironically, one of the arguments that apparently *was* used against Columbus by the advisers to the Spanish court *depended* on the assumption that the earth was spherical (combined with an understandable confusion about the workings of gravity). The argument advanced was that as one traveled farther and farther from home one would be sailing downhill at a steeper and steeper angle. Eventually one risked reaching a point where the uphill sail back home would be impossible, even with the strongest of winds.

It is hard today to appreciate the degree to which the earth and its oceans appeared immense and forbidding—the vast scale, the terrible dangers that explorers like Columbus were aware of, compounded with the even greater fears of the unknown. And all was magnified by rumors and tales enhanced in the telling. One extensive undertaking of the time—exploring the western coast of Africa—lasted almost an entire century. The exploration of the African coast by the Portuguese involved experts in navigation, cartographers, nautical instrument makers, and shipbuilders. During this period, generations of sailors, navigators, and captains saw firsthand during their voyages the changes in climate, the unfamiliar positions of the sun and the stars, and the appearance of whole new constellations, all brought about by the curvature of the earth.

The real issue for Columbus and his contemporaries

was not the shape of the earth, but its size; and on that question there was indeed room for controversy. During the five hundred years or so between Aristotle and Ptolemy there had been a number of estimates of the size of the earth. Eratosthenes' estimate was among those closest to the truth (although perhaps overestimating the correct value by about 10 percent, depending on how the units of measurement then used are interpreted today). Ptolemy chose for his *Geography* an estimate that was as much as 20 percent too low. Since by the fifteenth century Ptolemy was regarded as *the* authority on the earth's geography, the value he gave for the circumference of the earth—about 20,000 miles—was widely accepted. For Columbus, Ptolemy's estimate had the additional appeal of bolstering the explorer's case for the feasibility of sailing west to the Orient. (On an earlier trip to Africa, Columbus made measurements of his own that tended to confirm the figure given by Ptolemy.) In addition to underestimating the size of the earth, Ptolemy vastly overestimated the size of Asia. The resulting map depicted an earth whose expanse of ocean between the western tip of Europe and the eastern tip of Asia was well within the range of provisions that ships of the day could carry.

In 1484, King John II of Portugal appointed a team of experts, called the Junta dos Matemáticos, to review proposals for maritime exploration and to advise on their feasibility. The members of the Junta not only were well versed in all the scholarly work on geography and navigation but also had at their disposal the reports from the many earlier Portuguese voyages of exploration. It was undoubtedly their opinion—fully borne out by subsequent events—that Columbus was overly optimistic in his estimate of the distance to Asia.

After being rebuffed by Portugal, Columbus took his case to Spain, where—despite many delays—he finally received royal backing for his expedition.

Columbus was wrong not only in his estimate of the distance to the Orient but also in the assumption that there would be no land on his route between Europe and Asia. Since he was wrong on the first count—the distance was in fact far greater than his provisions could possibly have lasted—had he been right on the second and had a vast ocean to traverse, he would almost certainly have perished. To his very good fortune he was wrong on both counts, and in this case two wrongs made a triumphant "right," with all the attendant fame and glory.

Columbus' voyages literally changed the map of the world. Before those voyages, European mapmakers often depicted the world as a large circle or oval with Europe, Asia, and Africa inside, surrounded by ocean. Columbus would have rightly interpreted the left-hand border as depicting the same line on the earth as the right. In other words, sailing off the left-hand edge of the map would have resulted in sailing onto the right-hand edge of the map, just as, five hundred years later, video-game aficionados became accustomed to seeing images vanish from the left-hand edge of their screen and reappear on the right.

With Western civilization's discovery of the Americas, it became more convenient to picture the world in the form of two hemispheres—Eastern and Western—each depicted in a circle of its own. Sailing out of one circle on such a map is tantamount to sailing into the other one.

From the Eurocentric viewpoint of those making the maps, the Western Hemisphere became the "New World" and their own hemisphere, the "Old World."

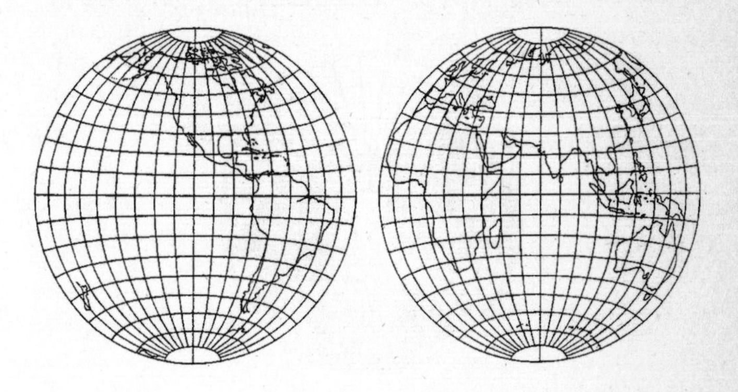

The Western and Eastern Hemispheres.

The depiction of the earth in the form of two hemispheres has become so commonplace that some of its subtler points may be overlooked. First, the division into Eastern and Western hemispheres is purely arbitrary—the earth can be divided in half for mapmaking purposes in any number of ways. A more natural division is into Northern and Southern hemispheres. In fact, any great circle on the earth could be used to divide the earth into two equal halves, each of which is depicted as the inside of a circle. In the case of the Northern and Southern hemispheres, their common boundary is the equator. Every point on the surface of the earth is either above the equator, below the equator, or on the equator. All points that lie above the equator are depicted on such a map as points inside one circle, while all points below the equator are depicted by points inside the other. A location on the earth lying right on the equator is depicted twice on the map—on each of the circles depicting the outer edges of the two hemispheres. For example, the point near the

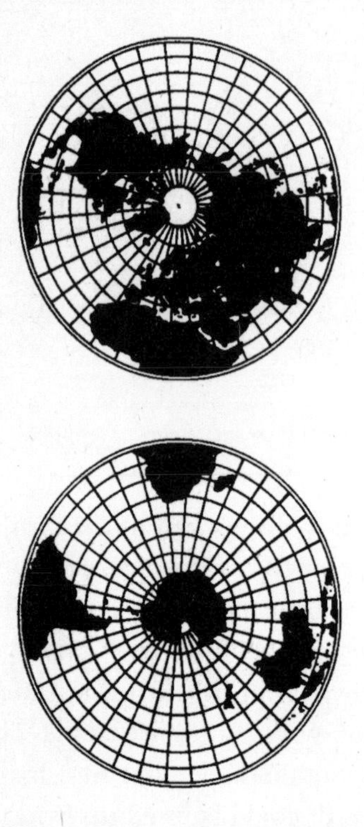

west coast of Africa where the Portuguese first crossed the equator appears on the outer edge of the Northern Hemisphere and on the outer edge of the Southern Hemisphere.

One of the less desirable features of the two-hemisphere map is that certain points that are close together on earth—say just above and just below the equator—may not be close at all on the map. It is, however, a drawback that cannot be overcome. One can prove, using a modern branch of mathematics known as *topology,* that *any* map of the surface of the entire earth must have the same weakness; one cannot concoct a way of depicting the whole earth on a sheet of paper so that nearby points on the earth will always be close to each other on the map.

This weakness, however, is more of a mild inconvenience than a serious problem to map readers and navigators. A more critical question is how one depicts the geographical features inside each hemisphere so that distances and directions can be reliably read off the map. Early maps were notoriously inaccurate. The incomplete and unreliable data upon which the maps were based were only part of the problem. More fundamental was the question of how to transfer measurements of latitude and longitude on the surface of the earth to corresponding locations on the map without producing vast distortions. For purposes of navigation it was particularly important to have a map on which one could rely. There are two features that navigators especially prize. First, that all routes due north from any point on the map be depicted by the direction "straight up." Second, that all compass directions be correctly depicted on the map relative to the northerly direction, so that a river flowing east-west would be horizontal on the map, and a road going northeast should be depicted at a 45° angle—halfway between horizontal and vertical.

Any map that has those two properties we will call a *navigator's map.* Such maps automatically have certain additional properties: parallels of latitude are depicted as horizontal lines, and the map has a fixed scale along each parallel of latitude—in other words, three points equally spaced along the same latitude appear equally spaced on the map. Most importantly, if one wants to sail from Lisbon to a particular point on the coast of North America, then setting a fixed compass direction indicated by the direction of the straight line between the corresponding points on the map provides a route to the precise destination.

The first actual map designed on these principles was drawn in 1569 by the Flemish cartographer Gerhard Kremer, whose last name means "merchant" and who went by its Latin equivalent: Mercator.

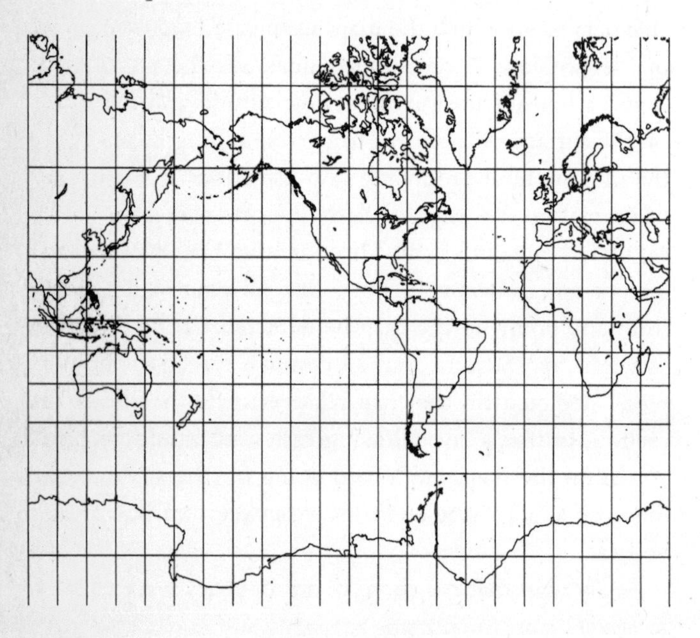

A Mercator projection.

Nowadays, the name Mercator is all but synonymous with this particular map. In his own time, however, Mercator's fame was not restricted to mapmaking. He was well known and enormously successful in the long tradition of master craftsmen in Renaissance Italy whose scientific instruments were equally prized as works of art and were collected by the Medici. In 1541, Mercator made a globe for Charles V, the Holy Roman Emperor, who then commissioned a set of surveyor's instruments. The combination of beauty and accuracy brought Mercator many commissions and considerable wealth and fame. But his primary passion appears to have been mapmaking; he spent many years working on a definitive edition of Ptolemy's *Geography,* and he was universally recognized as Europe's premier cartographer.

All of that meant little to posterity, which focused on the single map drawn by Mercator in 1569, of which a sole exemplar survives. Mercator has also been the victim of a quirk of language that has led to a widespread misconception about how his map is designed. The word "projection" is interpreted by cartographers in a much broader sense than in common usage or in mathematical terminology, where the expression conjures up the image of a small transparent globe, with all the desired geographic features drawn on its surface, and a cylinder wrapped around it, touching the sphere along the equator. A light placed at the center of the sphere will "project" the sphere onto the cylinder—that is to say, it will cast shadows of each continent onto the cylinder, and each meridian will project onto a vertical line. When the cylinder is cut along one of those lines and unrolled onto a plane, the result will look very much like Mercator's famous map, with many of the same features: meridians will be vertical lines, parallels of latitude will be projected as horizontal lines, and there will be equal scaling along each fixed latitude. Nevertheless, it is not a true Mercator map: all compass directions other than north-south and east-west will be wrong.

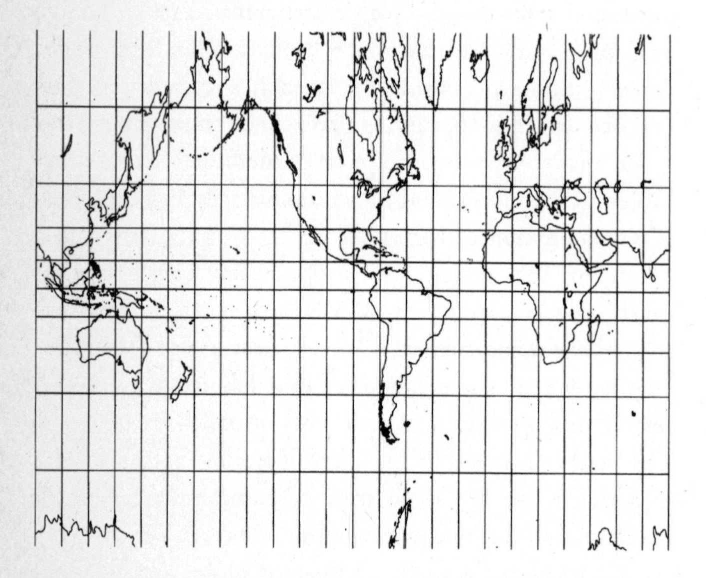

A cylindrical projection. The distortion of the map toward the poles is far greater for the cylindrical projection than for Mercator's map.

Mercator's map is not based on any such simple projection or geometric construction. Mercator explained the principles on which the map was drawn as follows:

> In making this representation of the world we have had to employ a new proportion and a new arrangement of the meridian with reference to the parallels . . . we have progressively increased the degrees of latitude towards each pole in proportion to the lengthening of the parallels with reference to the equator.

In other words, in order to have the angles come out right, Mercator stretched the map in the vertical direction (which is what he means by "increasing the degrees of latitude toward each pole") and the amount of vertical stretching is the same as the amount of horizontal stretching. But the amount of horizontal stretching along any parallel of latitude is simply equal to the ratio of the length of the equator to the length of the given parallel, since the equator and all the parallels of latitude appear as horizontal lines of fixed length: the width of the map. With these general guidelines and using as much art as science, Mercator produced his map. It was not until century's end that Edward Wright gave an explicit formula for the amount of stretching at any given latitude, and used it to compile a table of small increments of latitude and the corresponding locations on Mercator's map. Wright's tables allowed anyone to construct the map (without having to master the principles on which it was based). However, Wright's tables still provided only an approximation of a true Mercator's map. That was the best that one could hope for, since the exact equations for Mercator's map require the use of logarithms, which had not yet been invented at the time. It was not until 1668, just ninety-nine years after Mercator conceived the idea for his map, that mathematicians succeeded in applying the newly invented subject of integral calculus to obtain the exact equations for Mercator's map. With those equations in hand, anybody—or, in modern times, any computer—can produce a Mercator's map drawn to whatever degree of accuracy is desired.

The biggest drawback of Mercator's map was its progressive distortion from the equator toward the poles. A pair of points on a pair of vertical (or longitudinal) lines on the map remain a fixed distance apart, while the corresponding points on the actual surface of the earth get closer and closer together as they approach the pole. As a result, a region lying far to the north or south will appear much larger on the map than the same-size region located near the equator.

Ideally, one would like to have a map with the desirable features of Mercator's map but without the distortion. However, it is a geometrical fact (whose complete proof had to wait several hundred years) that the two properties of a navigator's map—a vertical north and directions relative to north depicted correctly—determine the map completely (up to the overall scale): it must be Mercator's map. Our choice is either to sacrifice one (or both) of the desirable properties or else to accept the distortion.

The problem is not limited to maps attempting to depict the entire earth. It arises with every map of a city, region, or country. Such maps typically provide an arrow that indicates the direction "north" for all points on the map, and a fixed scale, such as "an inch to a mile," meaning that the distance between any two points on the map, measured in inches, converts to the actual distance between the corresponding points on the earth, measured in miles. But it is impossible for any map to have both a fixed direction for north for all points on the map and a fixed scale. Such a map would automatically have to depict compass directions correctly, and thus would have to be Mercator's map; but it could not then have a fixed scale, since the scale on Mercator's map is different on different horizontal lines. The reason that a city or county map can *claim* a fixed scale and a fixed direction for north is that the variation of scale on a Mercator projection for a relatively small region (at least away from the poles) is small enough to be negligible.

Since a fixed scale on a map is incompatible with a fixed direction for north, one could ask if it is possible to produce a map of the earth with a fixed scale by allowing the northerly direction to vary from point to point, as it does in the usual depictions of the earth in two hemispheres. In other words, is there any map at all that has no distortions? Despite centuries of efforts leading to ingenious partial solutions to the problem, cartographers were continually frustrated; it was as if they were dealing with an unsightly tube of toothpaste—squeezing it at one place would always result in a bulge somewhere else. The problem was finally resolved in the mid-eighteenth century by the leading mathematician of the time, Leonhard Euler.

Euler's mathematical interests ranged from the purest theoretical investigations to the most practical applied problems. Intrigued by the mapmakers' difficulties, he proved conclusively that what they had been trying to ac-

Emanuel Handmann, Portrait of Leonhard Euler, *1756.*
(University of Basel, Museum of Natural Science)

complish was in fact impossible. There is no map of any portion of the earth's surface which, translated onto a flat sheet of paper, has a fixed scale. Every map, in fact, is a compromise.

As Euler's theorem shows, there can never be a perfect map. The challenge to cartographers is to devise new ways of mapping the earth that either minimize the overall distortion or else are particularly well adapted to a

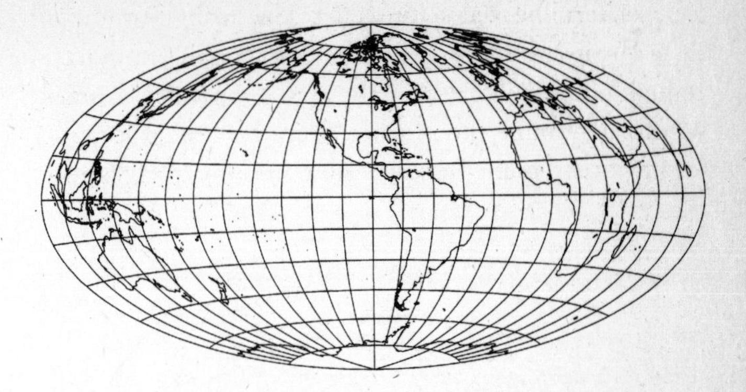

A Hammer projection. One of many attempts by cartographers to find new and better ways to depict the earth's surface. This one dates from 1892 and has been widely adopted.

certain purpose. They responded with literally hundreds of different maps, of which several dozen are in common use. For example, during the twentieth century, as ocean crossings were gradually replaced by air travel, the "navigator-type" maps became less relevant. Oddly enough, a much more useful map for navigating by air is one that had been invented long before Mercator's map. It was introduced by al-Biruni sometime around the year 1000, and is called an *azimuthal equidistant projection* by cartographers and an *exponential map* by mathematicians. A better name for it would be an *egocentric map.* Choosing one's hometown or other favorite location as the focus, one can draw such a map with that place at the center, surrounded by the rest of the world. Distances from the center to any other point on earth are drawn to scale, and directions around the central point are correctly indicated. These two properties, together with the given scale, determine the entire map. The main features of such a map are that it represents the region around the central point quite accurately, and it provides a quick way to determine the distance from the central location to any other point on earth: just measure the distance

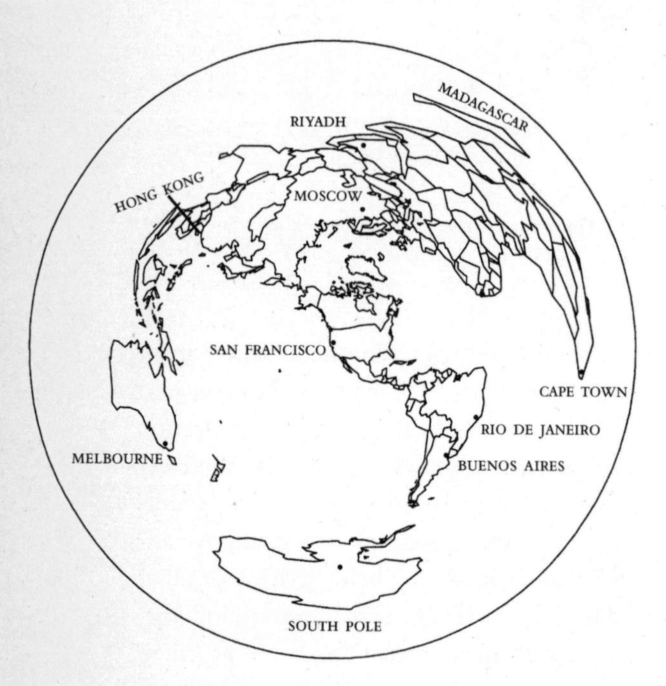

An egocentric map based at San Francisco.
The map shows at a glance that the direct route from San Francisco to Riyadh in Saudi Arabia goes directly over Moscow and Baghdad. (It also shows that Rio de Janeiro is farther from San Francisco than Buenos Aires and only a little closer than Hong Kong.)
Since the antipodal point to San Francisco lies in the Indian Ocean, off the coast of Madagascar, the whole outer ring of the map corresponds to a small piece of the Indian Ocean surrounding the antipodal point, and Madagascar is stretched out on the map to roughly the same length as North America.

on the map and multiply it by the scale factor. Furthermore, the straight line on the map joining the center to any other point shows instantly which cities, countries, and other landmarks one will pass over if one flies the direct route from the central location to the second point.

As Euler's theorem tells us, these attractive features of an egocentric map relative to its central point must necessarily be paid for by distortions elsewhere on the globe. And in fact, the further away from the center that one looks, the more the map is distorted. The reason is that each circle around the center of the map corresponds to a circle of points on the globe at a given distance from the central location. As those distances approach half the circumference of the earth, the circle on the map gets larger, since it is further from the central

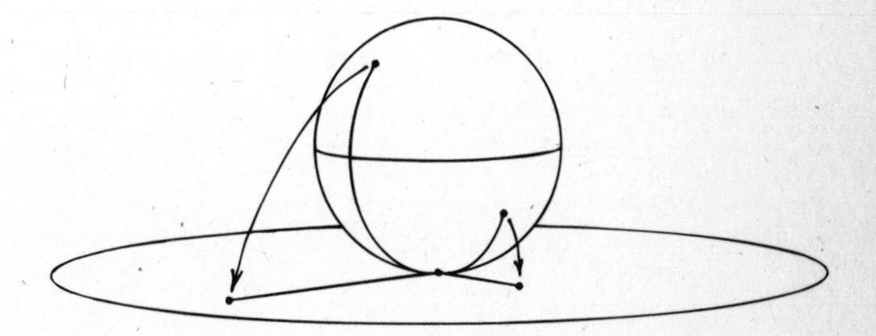

How to construct an egocentric map based on your hometown: *Place a globe of the earth on a large sheet of paper with your hometown's location on the globe touching the paper. To find the precise point on the map corresponding to any position on earth, stretch a string along the globe from the point of the globe touching the paper (your hometown) to the desired position on the globe, and then stretch the string along the line of the paper directly below where it was on the globe.*

point, while the corresponding circle on the globe contracts down toward the antipodal point: the point on the earth exactly halfway around from the central location. The distortion is most extreme at the antipodal point, which is not represented by a single point on the map, but is stretched out to an entire circle—the outer boundary of the map. In fact, starting at any point on earth and heading off in any direction always leads to the antipodal point after going a fixed distance: halfway around the globe.

Little did al-Biruni suspect that his new map of the world would someday—many centuries later—prove singularly useful for future modes of navigation. Even less could he have foreseen that his map would someday provide a particularly appropriate way to view and understand the entire universe. The best route to that understanding is not the direct one, however, and we shall have to make several side excursions along the way to explore the key notion of "curvature" before arriving at our destination.

CHAPTER III

The Real World

AN ELEGANTLY EXECUTED PROOF IS A POEM IN ALL BUT THE FORM IN WHICH IT IS WRITTEN.

—Morris Kline,
author of *Mathematics in Western Culture*

Carl Friedrich Gauss and Ludwig van Beethoven led parallel lives. Born less than seven years and two hundred kilometers apart, they came to epitomize the height of their respective professions, mathematics and music. Presumably, true to the nature of parallels, they never met. They both acquired an almost superhuman aura among their contemporaries and for subsequent generations. During his lifetime Gauss earned the semi-official title of *princeps mathematicorum,* "first among mathematicians."

Gauss's singular achievements were by no means limited to mathematics. He may well have been the last of the great all-around scientists; in the tradition of Newton, he made profound contributions across the spectrum of pure and applied mathematics as well as in physics and astronomy.

In the field of physics, Gauss devoted many years to the study of electricity and magnetism. In addition to a number of theoretical contributions, he carried out ex-

Carl Friedrich Gauss and Ludwig van Beethoven at around thirty years of age. (Gauss: Universitäts-Sternwarte Göttingen; Beethoven: Beethoven–Haus Bonn, H. C. Bodmer Collection)

tensive experiments toward determining the strength of the earth's magnetic field. To that end, he developed an absolute scale for measuring magnetic fields. As a result, the standard unit of magnetic field strength is called a "gauss." As a practical application of electricity and magnetism, Gauss and his collaborator, Wilhelm Weber, invented a telegraph that they used in the 1830s to communicate between the observatory and the laboratory at Göttingen—a distance of about a mile. (Samuel Morse obtained his patent on the telegraph in 1840.)

Astronomy was the domain in which Gauss first gained worldwide recognition. Ceres, the first asteroid to be discovered, was originally spotted during the night of

January 1, 1801, by an Italian astronomer, Giuseppe Piazzi, who followed its nightly progression through early February, when it disappeared on its way around the sun. Young Gauss, then twenty-four, was one of a number of scientists who used the data given by Piazzi to calculate where Ceres was likely to reappear toward the end of the year. His prediction turned out to be remarkably accurate, and led to the first sightings of Ceres when it was again visible. In the process, Gauss developed new methods in probability and statistics, including a study of the famous "bell-shaped curve"—now known as the "error curve" or "Gaussian distribution"—which plays a central role in the analysis of data of all kinds.

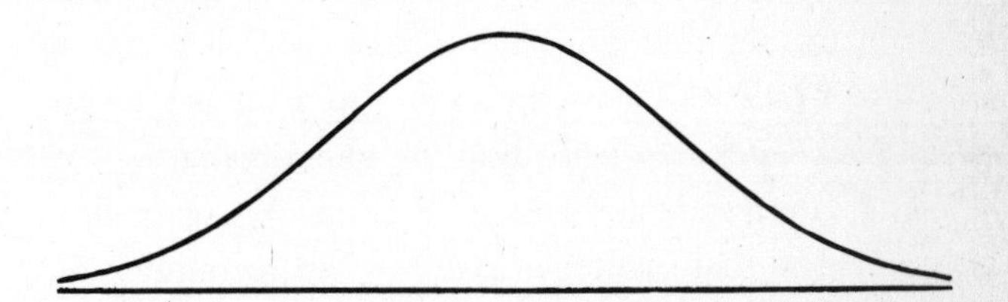

A "bell-shaped curve" or "Gaussian distribution"—ubiquitous in probability and statistics.

Following his initial foray into astronomy, Gauss immersed himself more and more in the subject. He was interested not only in the theoretical aspects of the field but also in making observations and in the design of telescopes. His growing reputation as an astronomer led to an offer, at age twenty-nine, to head the observatory at Göttingen. He accepted, and in 1807 moved to Göttingen, where he was to remain for the rest of his life.

Gauss spent a considerable part of his middle years on a study much less glamorous than the astronomical research of his early years, or his later investigations into magnetism; nonetheless, Gauss's intervention in this study was to have even more far-reaching consequences. In 1818, Gauss agreed to become director of a large-scale project to carry out a complete survey of the kingdom of Hannover. In typical fashion, Gauss entered with equal energy into both the practical and theoretical sides of the project. He went out into the field to personally direct and participate in surveying, and he retired to his study to devise new methods to handle and interpret the data.

What makes Gauss a truly unique figure in the world of science and mathematics, standing out even in comparison with other towering talents such as Euler, was his ability to penetrate below the surface of a subject, to uncover the deeper reasons behind the phenomena.

The quintessential Gauss is revealed in a story from his childhood that mathematicians love to recount, as did Gauss himself in his later years. It describes a triumphal moment in elementary school, when his teacher asked the class to add up all the numbers from one to a hundred. Gauss simply wrote down the answer: 5050, and sat patiently while his classmates laboriously carried out their computations. Gauss noticed that by pairing the first and last numbers, the second and next-to-last, and so on, each pair added up to a hundred and one. The numbers from one to a hundred fall into fifty such pairs and so the total sum is fifty times a hundred and one, or 5050.

What is the appeal of this story? In large part it is the victory of ingenuity over drudgery (especially teacher-

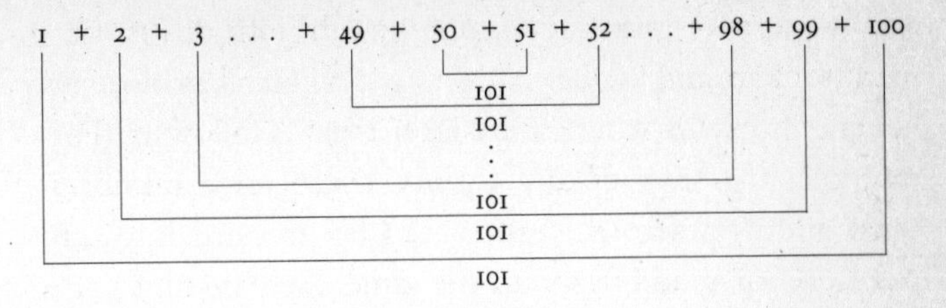

Pairing numbers from one to a hundred.

imposed drudgery). It is also emblematic of what mathematicians call an "elegant" solution to a problem. It illustrates strikingly the degree to which what matters in mathematics, as elsewhere, is not just getting the answer but how you get it. (In the words of an old song: "It ain't what you do, it's the way that you do it; that's what gets results.") When Sherlock Holmes solves a mystery, it is not simply *that* he solves it but *how* he does it that keeps us transfixed. Exactly the same is true in mathematics. Anyone can add the numbers from one to a hundred and get the answer; Gauss obtained the answer *without* adding them.

Beyond these features, what the story reveals is how a deeper understanding of a problem not only can lead more quickly to an answer, but can also explain why the answer takes the particular form that it does. Adding up the numbers may give the correct value of 5050, but it will provide no clue, for example, why the answer ends in a zero. The underlying reason—that the sum of the numbers from one to a hundred is equal to a product: half the number of terms (the fifty pairs) times the sum of the first and last terms—gives much more insight into this particular problem, and applies equally to a whole class of more general problems where a similar pattern is followed. (For example, the sum of any twenty consecutive numbers must end in a zero since they can be grouped into ten pairs, each pair adding up to the same fixed amount.)

Carrying out a large-scale land survey, such as the one Gauss agreed to head, has certain aspects in common with performing a long series of additions: it is a relatively straightforward, tedious, time-consuming, and error-prone operation. It bears closer resemblance to adding a set of random numbers than to adding an orderly succession of numbers, in that no shortcuts are available to cut down the work. Nevertheless, Gauss once again used a seemingly mindless task as a springboard to a line of reasoning that was to have profound consequences.

In order to understand Gauss's idea it is worth examining *geodesy*—the theoretical study underlying large-scale surveying—in a bit more detail. The standard procedure used in carrying out a geodetic survey is called a "triangulation." A number of landmarks are selected, and distances between different pairs of landmarks are carefully measured. In that way the region surveyed is covered by a network of triangles whose sides and angles have been determined as accurately as possible. From that information one can infer other measurements, such as the "distance as the crow flies" between two distant landmarks. However, the way the triangles fit together depends on the size and shape of the earth. If the earth were flat, the standard formulas of euclidean geometry would apply. If the earth were a perfect sphere, one could use *spherical geometry*—

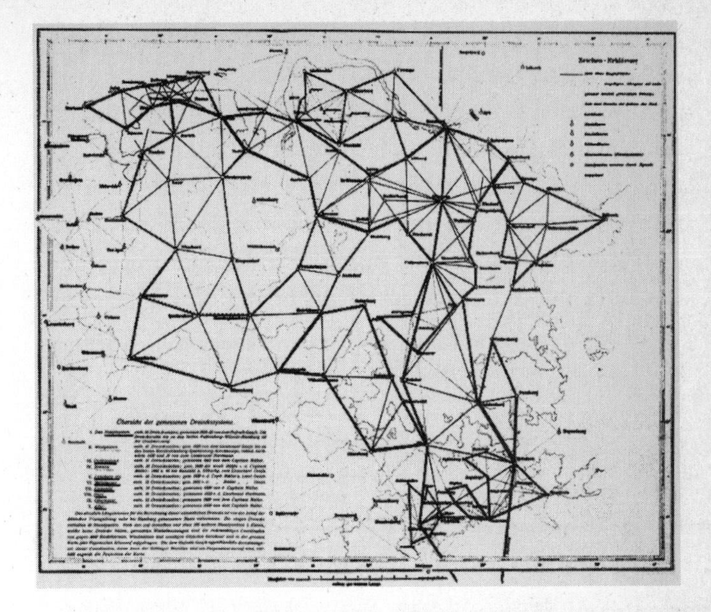

Triangulation of a region extending from near Göttingen on the south to Hamburg on the north and roughly the same distance east to west. This was the geodetic survey carried out by Gauss during the period 1821 to 1838.

the geometry of figures on the surface of a sphere. In reality, the earth is neither flat nor perfectly spherical. In addition to the unevenness caused by mountains and valleys, there is a more sizable deviation from sphericity due to the earth's rotation. In fact, Isaac Newton concluded that the shape of the earth should be slightly ellipsoidal, bulging at the equator and flattened at the poles. Using the same equations—"Newton's laws"—from which he had derived the motions of the planets, Newton was able to calculate the size of the bulge at the equator. (Later surveys confirmed Newton's predictions, showing that the earth's circumference is 24,902 miles around the equator and only 24,860 miles over the poles.)

One effect of the earth's ellipticity is to skew the results of measurements, such as those of Eratosthenes', of the size of the earth. For an elliptical cross section, the directions "straight up" do not change uniformly. For example, if one travels five hundred miles due north and measures the difference in height of the North Star, the

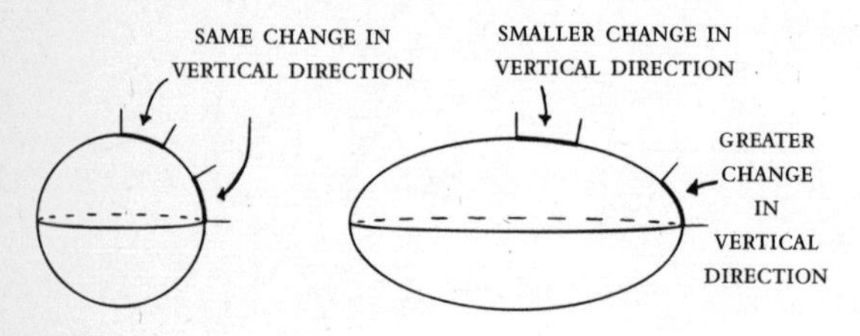

How the ellipticity of the earth affects geographical and astronomical measurements. On a spherical earth, the vertical direction changes by a fixed amount as one travels a fixed distance due north, no matter where one starts. On an elliptical earth, the vertical direction changes less along the flattened part of the earth near the poles than along the more curved part near the equator. (The amount of flattening in the picture is much greater than for the earth, but the same principle applies.)

answer would depend on where one started; there would be greater variation near the equator and less toward the poles. In fact, careful observations and measurements of just that type can be used to determine the shape of the earth. In turn, knowledge of the overall shape of the earth is critical in order to interpret correctly the data obtained from a geodetic survey. One of Gauss's contributions to geodesy was perfecting the mathematical tools needed for that purpose. But Gauss's most profound insights came from viewing the problem in reverse: not how the shape of the earth affects the results of a survey, but how a survey can be used to determine the shape of the earth. Suppose, for example, that the climate on earth were somewhat different, so that we had—like Venus—a perpetual cloud cover. We could not then use the sun and the stars, or eclipses of the moon, as props to help determine the shape of the earth. Could we then determine if the earth was round or flat, spherical or elliptical, by just making geodetic-type measurements along the earth's surface?

Gauss's answer was "yes." While we cannot determine the shape completely, there is a surprising amount that we can deduce. For example, from such measurements on the surface alone, we can easily confirm that the earth cannot be flat and that it is more elliptical than spherical.

To see why that is the case, picture the process of planting a large orchard. We might start with a long rope, knotted at equal intervals, indicating the ideal spacing between the trees. We could stretch the rope along the ground, pulling it tight to make it as straight as possible, and then we would plant a tree at each knot. If we wanted to extend the line of trees past the original length of rope, we could shift the rope a few notches along its original line, adding as many more trees as we liked, with the same spacing, along the same line. The next step would be to plant columns of trees. Starting at each tree of our first row, we could stretch the rope in the direction perpendicular to the original line of trees, and again place a tree at each knot. If we draw a sketch in which the trees of the original row lie on a horizontal line, and the perpendicu-

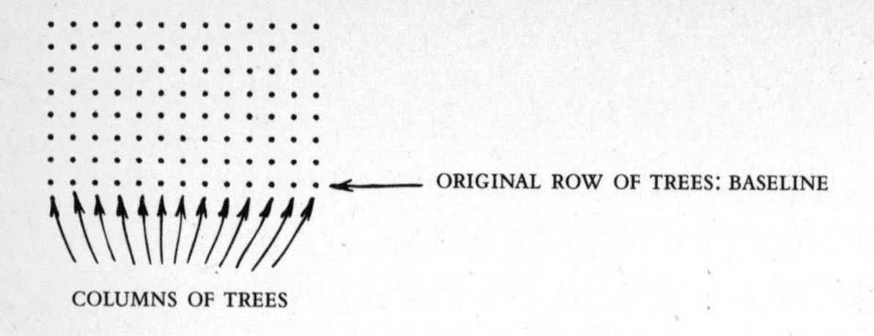

An "orchard" in a plane.

lar columns form vertical lines, then we will see the appearance of a whole series of horizontal rows, seemingly parallel to the original row. (Anyone who has walked or driven past an orchard is likely to have been struck by the appearance of many different diagonal rows of trees, looking as if they too had been laid out in straight lines.) If the earth were truly flat, euclidean plane geometry would apply perfectly: not only would the horizontal rows of trees lie along straight lines parallel to the baseline, but the trees on each of those rows would again be equally spaced, with the *same* spacing as in the original row of trees. Because relatively small portions of the earth *are* very close to being flat—the same fact that gave rise to euclidean geometry in the first place—the spacing along the horizontal rows *is* very close to that of the original baseline, and it has the practical consequence that farmers can use machines specifically designed to travel between the columns of trees on the premise that the width between the columns stays fixed.

But imagine an orchard large enough so that the deviation from flatness is easily perceptible. Suppose that our original tree line were along the equator, and that we had a truly large plantation, stretching east-west several degrees of longitude. Then the vertical columns of trees would be planted along equally spaced meridians. The resulting horizontal rows of trees, as one traveled north, would appear initially to have the same spacing as the original row along the equator, but if the plantation were sufficiently large, the compression of the meridians toward the poles would have a measurable effect on the spacing between the trees.

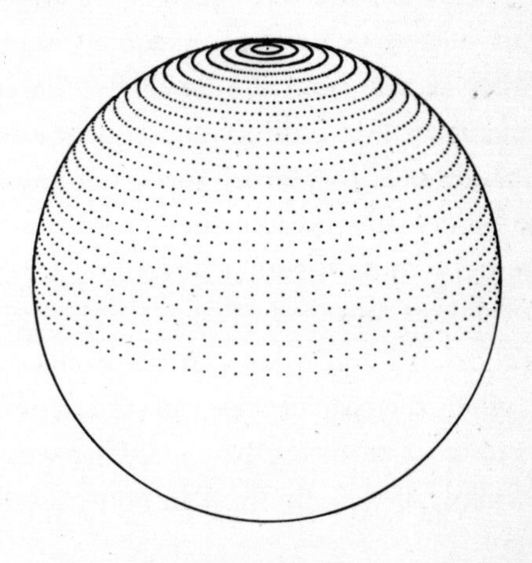

An "orchard" on a sphere: positive curvature.

In order to develop formulas for the variation in the spacing of meridians from the equator toward the poles—or, in the example of the larger orchard we have just imagined, formulas for the change in spacing between trees in the horizontal rows as their distance from the equator increases—geographers had to replace euclidean plane geometry with a newer, more advanced branch of mathematics: *spherical geometry.* But once again, the new formulas were accurate only to a degree, since the earth is not a smooth sphere, but a bumpy ellipsoid. Gauss's contribution was a set of formulas that may be used on *any* surface at all, whether a plane, a sphere, an ellipsoid, or a completely general surface. These formulas, applied to our example of the large orchard, relate the spacing between the trees in successive rows to a quantity now known as the *Gauss curvature,* or simply the *curvature,* assigned to each point of the domain to be planted or surveyed. That the trees in successive rows get gradually squeezed together as they recede from the baseline—the equator in the example above—is a direct consequence of the fact that the curvature of the earth's surface at any point is a positive number—the greater the curvature, the more rapidly the trees come together, according to an explicit formula given by Gauss.

Gauss's formula applies also in the event of negative curvature, when the spacing between the trees would *increase* with distance from the baseline. That would be the case, for example, if we colonized an hourglass-shaped as-

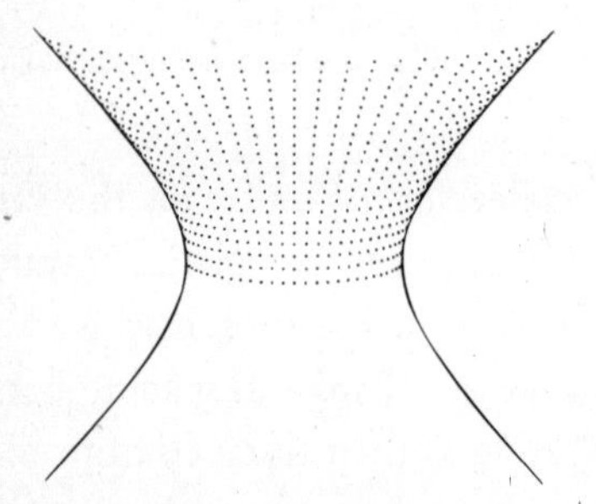

An "orchard" on a hyperboloid: negative curvature.

teroid, and placed our plantation along *its* equator. Again, the more negative the curvature, the more rapidly the spacing would increase. Only in the case of zero curvature, as on a flat plane, would the trees remain equally spaced.

Gauss curvature played such a seminal role, not only in studying the shape of a surface but also in subsequent attempts to understand and picture the universe, that it is worth closer examination from several different points of view.

There is one aspect of Gauss curvature that invariably causes confusion on first exposure. A surface such as a cylinder certainly appears to be curved, but nevertheless its curvature (in the sense of Gauss) is *zero*. The reason is that as far as measurements *along the surface* go, there is no way to distinguish a part of the cylinder from a part of the plane; one can just roll up a rectangle in the plane to form a cylinder, without any stretching or distortion. No survey

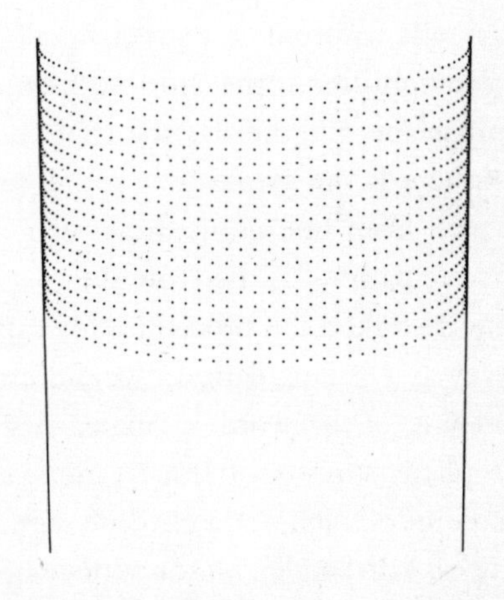

An "orchard" on a cylinder: zero curvature.

made on the surface would reveal the difference (unless we go far enough around the cylinder to end up where we started). In terms of our orchard-planting model, if we started planting around a "waist" and then made successive rows, the spacing would remain the same, which is the telltale sign of zero curvature.

Let us look more closely at the actual process of measurement involved in carrying out a geodetic survey. Central to the process is the notion referred to popularly as "distance as the crow flies." How does the crow fly? The "crow" in this expression veers neither right nor left, but follows a path that always goes "straight ahead." Any such path along a surface—the crow's flight path or "straight-ahead path" along the surface—is called a *geodesic.* On a plane, geodesics are straight lines; on the surface of a sphere, they are great circles, such as the equator and the meridians. When we stretched out our rope to make it as tight as possible before planting our first row of trees in the orchard, it was to produce a geodesic along the surface. Just as a straight line is the shortest distance between two points in a plane, and the shortest route between two points on a sphere is the great circle route, so in general the shortest path from one point to another on any surface is a geodesic path—"as the crow flies."

When we described the procedure for carrying out a geodetic survey as a "triangulation," we did not make explicit the meaning of the word "triangle." A standard triangle in the plane consists of three points, called "vertices," and three sides, each a straight-line segment joining a pair of vertices. A "triangle" on the surface of a sphere—also called a "spherical triangle"—consists of three vertices joined pairwise by arcs of great circles. On an ellipsoid, such as the surface of the earth, or on any other surface, a "triangle" is understood to mean a figure consisting of three points—again called "vertices"—joined by geodesic arcs, called the "sides" of the triangle. (The more formal name for such a figure is a "geodesic triangle.")

A fundamental fact about triangles in the plane is that the sum of the angles at the three vertices is the same for all triangles: 180°. For triangles on a sphere, the sum of the angles turns out to be always *greater* than 180°, and, furthermore, to depend on the size of the triangle. For small triangles, the sum of the angles is only slightly greater than 180°, but there are *large* equilateral triangles with three right angles; one such example can be seen by taking an arc of the equator going one quarter of the way around, and joining each end to the North Pole by a meridian.

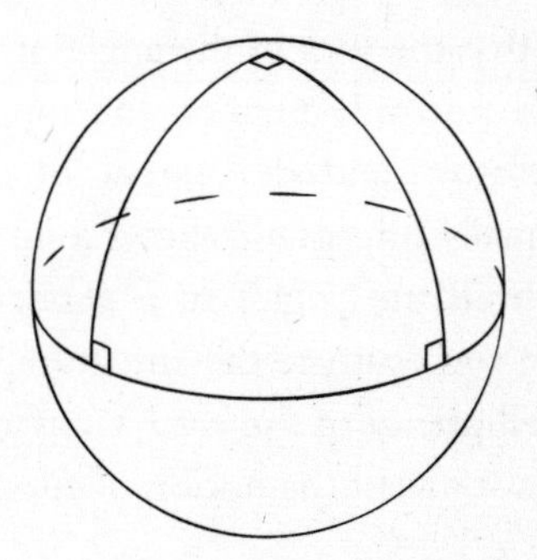

A three right-angled triangle on a sphere.

Suppose that the earth were perfectly spherical and we wanted to determine its size without resorting to extraterrestrial aids, such as the stars or the sun. As it turns out, the *only* size equilateral triangle on a sphere with three right angles is one in which each side is a quarter of a great circle. So the full distance around—the quantity Eratosthenes calculated—is just four times the length of the sides of such a triangle. That gives a theoretical answer to the question of how to determine the size of a sphere using just measurements along its surface. In practice one can use much smaller triangles and more complicated formulas to determine the size of a sphere. What Gauss was able to do was to find a formula for the sum of the angles of a geodesic triangle on *any* surface. The gist of the formula is that the sum of the angles depends on two factors: the size of the triangle and the value of the curvature at each point inside the triangle. When the Gauss curvature is zero—as for a plane triangle—the sum of the angles is 180°, independent of the size of the triangle. Positive curvature leads to a sum greater than 180°, and negative curvature ensures that the sum of the angles is less than 180°. Gauss's formula not only predicts the sum of the angles when the curvature and the extent of the triangle is known, but, equally important, allows a surveyor who has carefully measured the angles in a geodesic triangle to work in reverse and estimate the curvature, and from that, the degree of ellipticity of the earth's surface.

Gauss did not invent the notion of curvature that now

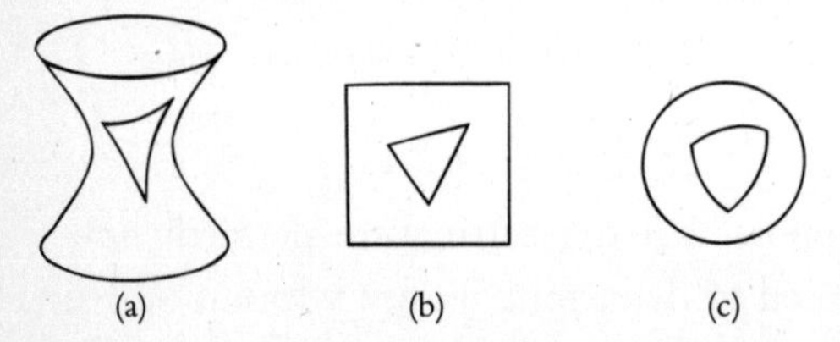

Triangles on three different surfaces: (a) a surface with negative curvature, (b) a surface with zero curvature, and (c) a surface with positive curvature.

bears his name. Other mathematicians before him had studied the same quantity. Gauss's important discovery—one that fully merits attaching his name to the concept—was that this curvature is one of the quantities that can be determined by a geodetic survey—just by making measurements on the surface.

One of the many consequences of Gauss's work was a new way of proving Euler's theorem that a perfectly scaled map of any portion of the earth's surface is impossible. If there *were* such a map, then by scaling the earth down by the same factor, all measurements on the surface of the scaled-down earth would be exactly the same as on the map. That means that all quantities determined from such measurements, including the Gauss curvature, would be the same at each point on the scaled-down earth as at the corresponding point on the map. But the map, being drawn on a flat plane, has Gauss curvature zero, whereas the earth—whether a sphere or an ellipsoid—has positive Gauss curvature. So such a map is not possible.

Gauss's approach went beyond Euler's in several important ways. First of all, it revealed the underlying geometric meaning of the argument that Euler had given for a sphere; second, it was far more general, able to be applied to the real—roughly ellipsoidal—surface of the earth itself; third, it developed a new notion of curvature that was to take on increasing importance as time went on.

Gauss's characterization of curvature in terms of geodesic triangles has since been supplemented by others. Two French mathematicians, Joseph Bertrand and Victor Puiseux, gave a description that goes back to the basic notion of a circle. What does one mean by a "circle" of given center and radius on a sphere, ellipsoid, or other surface?

In the plane, a circle is the set of all points at a given distance (the radius) from a fixed point (the center). Exactly the same definition works on any surface, if by "distance" we understand distance "as the crow flies"—that is, along a geodesic. For a sphere, we would mean the shortest distance along the surface, or the great circle distance. If the earth were a sphere with a circumference of 24,000 miles, then the distance from the North Pole to any point on the equator would be exactly 6,000 miles; so the equator would be precisely the circle of radius 6,000 miles with center at the North Pole. The circumference of this "circle"—that is, the length of the equator—is 24,000 miles, or exactly four times the radius. This is much less than on a plane, where the circumference of any circle is more than six times the radius ($2\pi r$). In fact, every circle on a sphere has a circumference that comes up short compared with a circle of the same radius in the plane (where "radius" on the sphere always means "as the crow flies"). The *amount* that it is short precisely determines the curvature

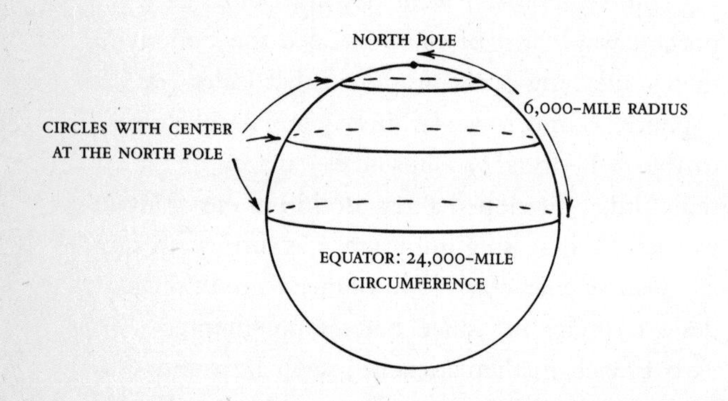

The equator as a circle on the earth centered at the North Pole.

of the sphere. Bertrand and Puiseux observed that the same is true on any surface. The circumference will fall short of the value for a circle of the same radius in the

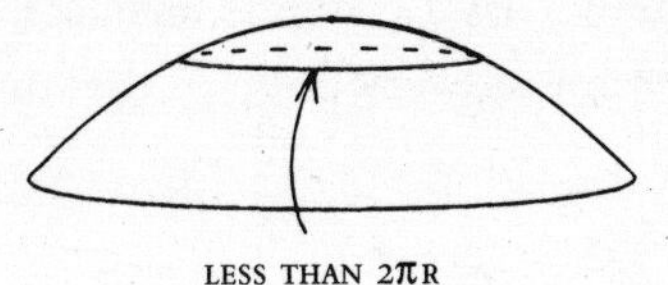

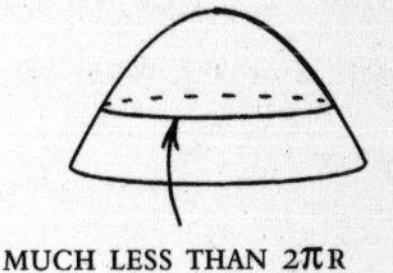

The length of circles on a surface with positive curvature.

plane whenever the curvature of the surface is positive; the greater the curvature, the greater the shortfall. If the circumference is larger than on a plane circle of the same radius, then the surface has *negative* curvature; the larger the ratio between the circumference and the radius, the more negative the surface's curvature.

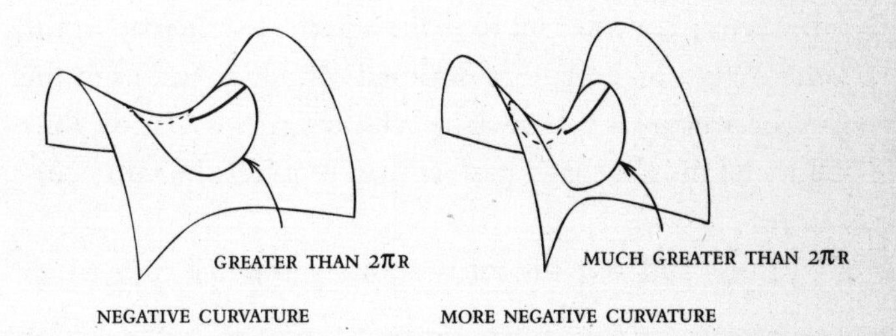

The length of circles on a surface with negative curvature.

A famous example of a surface with negative curvature is the inside-out sphere called a "pseudosphere." It looks like a drawn-out trumpet bell. It has not only negative curvature but *constant* negative curvature. That may come as something of a surprise, since different parts look very different. However, two circles of the same radius on a pseudosphere will always have the same circumference, no matter where they are drawn. By Bertrand's characterization, the curvature is the same at every point of the sur-

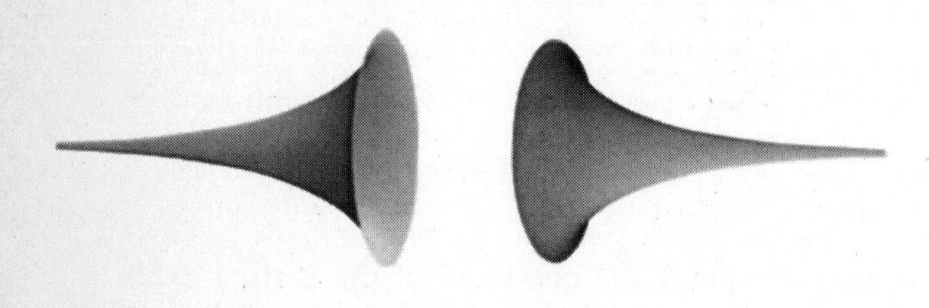

Two views of a pseudosphere.

face. It follows that if a design is drawn anywhere on the pseudosphere, an exact map of it can be made anywhere else on the pseudosphere, without distortion. On the other hand, any attempt to make a perfect scale map of the same design on a plane is doomed to failure, just as in the case of a normal sphere. The reason again is Gauss's fundamental insight that a perfect map requires the same curvature.

Gauss laid out his ideas about geometry in a paper written in 1827. The previous year saw the birth of a mathematician who was destined to extend Gauss's ideas in undreamed-of directions. But before coming to him we must describe an event that took place in 1829 that was to prove a turning point in geometry, with ramifications far beyond the world of mathematics.

CHAPTER IV

Imaginary Worlds

AS THE GEOMETER WHO SETS HIMSELF
TO SQUARE THE CIRCLE AND WHO CANNOT FIND,
FOR ALL HIS THOUGHT, THE PRINCIPLE HE NEEDS,

JUST SO WAS I ON SEEING THIS NEW VISION.
I WANTED TO SEE HOW OUR IMAGE FUSES
INTO THE CIRCLE AND FINDS ITS PLACE IN IT . . .

HERE POWERS FAILED MY HIGH IMAGINATION.

—Dante,
The Divine Comedy

One theme that recurs time and again in the history of mathematics is the gradual evolution of a new concept—from its initial rejection as being too abstract, through grudging acceptance of its usefulness, despite the fact that it appears "unnatural" and counterintuitive, to its eventual elevation to the status of a basic and indispensable tool in applications. One such example is the notion of "negative numbers." For centuries that expression was viewed as an oxymoron, a self-canceling phrase, a numerical absurdity; numbers count or measure things—there is no shape that has negative area, no circle with negative circumference, no book with a negative number of pages. For literally hundreds of years, great pains were taken to solve problems by methods that circumvented the use of negative numbers. Only very gradually did it become clear that the effort spent in avoiding their use was wasted effort, for negative numbers, though not interpretable in the same fashion as positive numbers, were just as acceptable, and in no way contradictory.

The notion of an imaginary number—one whose square is a negative number—followed a similar pattern of initial rejection and gradual acceptance. The problem was that the ordinary rules of arithmetic tell us that the product of two positive numbers is positive and the product of two negative numbers is also positive. As a result, any number multiplied by itself produces a positive number (or zero, if the original number is zero), never a negative number, such as -1. However, it turned out to be very convenient to act *as if* there were a number whose square was -1. The letter "i" was used to denote this new entity, and it was called an "imaginary number." It had the one peculiar property that its square was -1, but otherwise was subject to all the usual rules of arithmetic. The introduction of such a new species of "number" was both an imaginative and a risky act, since there was the possibility that the use of imaginary numbers could become more and more commonplace, and only much later lead to a serious contradiction, in which case all the previous work would have to be discarded. However, by the nineteenth century, when number systems were examined much more closely, it became clear that "imaginary numbers" were no more or less "real" than the standard "real numbers." Both are mathematical abstractions, and the "real numbers" include not only negative numbers that had been looked upon for so long with great suspicion but also oddities such as infinite decimals that never repeat and that do not satisfy any algebraic equation. And so imaginary numbers came to be fully accepted as part of a mathematician's tool kit, available whenever needed to solve problems. Imaginary numbers are now routinely used by engineers and physicists, and many applications of mathematics would be unthinkable without them.

But few developments in mathematics met with as much resistance and even outrage as that of non-euclidean geometry. By the nineteenth century, Euclid's geometry was two thousand years old, and had been a central component of a general education for centuries; it was also the prototype of clear thinking and logical reasoning. Immanuel Kant considered euclidean geometry to be hard-wired into our brains, the very essence of how the outside world is viewed and made viewable by each of us.

Yet in 1829, a Russian mathematician, Nikolai Ivanovich Lobachevsky, published a paper in which he presented an alternative to euclidean geometry. Many statements from euclidean geometry remained true in Lobachevsky's geometry: base angles of an isosceles triangle are equal, the largest side of a triangle is opposite the largest angle, and so on. In fact, the statements and proofs of the first twenty-eight propositions in Euclid's *Elements* hold without change in Lobachevsky's geometry. However, some of the most familiar theorems from euclidean geometry are no longer true: the Pythagorean theorem and the fact that the sum of the angles in any triangle is 180°, for example. In Lobachevsky's geometry, the sum of the angles in a triangle is not a fixed value, but depends on the triangle; in all cases, the sum is *less* than 180°.

Lobachevsky did not call his geometry "non-euclidean"; he termed it "imaginary." The reason Lobachevsky used the term "imaginary" was not that he considered his geometry any less real than euclidean geometry, but because many expressions in spherical geometry had counterparts in his geometry that could be obtained by simply substituting imaginary numbers in place of real ones.

Lobachevsky's work was greeted initially with very little response, due in part, perhaps, to his choice of the word "imaginary" to describe his geometry and in part to the fact that he published his paper in a fairly obscure journal, and in Russian. But what reaction there was, first by Russians who read his work, and then by other nationalities after it was translated, was almost uniformly negative.

Even more devastating was the experience of a young Hungarian named János Bolyai, who independently discovered non-euclidean geometry. Bolyai's father, Wolfgang, was a lifelong friend of Gauss. They met as fellow students in Göttingen, and after Wolfgang returned to Hungary in 1799, they stayed in contact, corresponding for over fifty years. Wolfgang, rightly proud of what his son accomplished, sent a copy of his manuscript to Gauss. A word of praise from Gauss, and best of all, a mention of Bolyai's work to Gauss's many mathematical contacts, would have launched the young Bolyai on what might have been a brilliant mathematical career. Instead, Gauss on this occasion revealed a less admirable side of his character. In his famous response to Wolfgang he explained that he could not praise the work, for to do so would be to praise himself. The reason was that Gauss had in fact anticipated much of what Lobachevsky and Bolyai had done. He had never published his own work, not because it had not yet reached the level of completion that his own high standards would deem satisfactory, but rather because he feared the negative reception he anticipated would greet it. Given the circumstances, he could just as easily have praised young Bolyai and encouraged him on. As it was, Bolyai was crushed to discover that the enormous effort he had expended in working out the details of the new geometry was simply wasted in retracing steps Gauss had taken many years before.

The irony of the situation is that Gauss was one of the few mathematicians who really did understand the new geometry and appreciate its importance. In fact, Gauss undertook at age sixty-two to learn Russian—partly because he delighted in learning new languages, partly as a challenge to assure himself that his mental powers were still intact, and partly, it appears, for the specific purpose of reading Lobachevsky's work.

Most of the mathematical world had two major doubts to resolve about non-euclidean geometry. The first was whether or not the new geometry was a viable candidate as an alternative to euclidean geometry. The second was whether it was of any value, if it was indeed "viable." Bolyai and Lobachevsky (and Gauss) had developed their new geometry by starting with the axioms of euclidean geometry, and replacing one of them—the "parallel axiom"—with a new axiom in direct contradiction to the euclidean axiom. All theorems of euclidean geometry that did not use the parallel axiom would automatically be true in the new geometry. However, theorems that used the parallel axiom would now be replaced by other statements which in the context of euclidean geometry seemed absurd. The big question—that of viability—was whether or not the new set of axioms would eventually lead to a contradiction, in which case the whole edifice was worthless, except insofar as it gave further credence to the belief that Euclid's geometry was the only possible geometry. Bolyai and Lobachevsky had pushed far enough with their new set of axioms to convince themselves that what they had was indeed a viable geometry and that the new set of axioms would not lead to a contradiction; however, they had no proof of that, and so doubts remained.

As to the *value* of the new geometry, assuming it was free of contradictions, Lobachevsky was well aware that it posed a fundamental question for all of science. Was the space we live in really euclidean, as everyone assumed, or might it be that the correct description of the real world was provided by Lobachevsky's "imaginary geometry"? There is a widespread but erroneous belief that Gauss carried out an experiment to see if he could decide whether space was euclidean or Lobachevskian by measuring the angles of a large triangle formed by three mountain peaks. Gauss did in fact carry out such an experiment, but it was in connection with a question in geodesy, and not as a means of establishing the euclidean or non-euclidean nature of space.

After the publication of his initial paper on the new geometry, Lobachevsky continued to develop his ideas. He published a second paper, entitled "Imaginary Geometry," in Russian; a French version appeared in 1837 in *Crelle's Journal,* one of the leading mathematical periodicals of Europe. Two years later, in the same journal, the German mathematician Ferdinand Minding introduced for the first time the surface that we now call the *pseudosphere.* In 1840, again in *Crelle's Journal,* Minding noted a remarkable fact: if one takes the standard formulas relating the lengths of the sides and the angles in a spherical triangle, and if one substitutes an imaginary number for the radius of the sphere, then the resulting formulas are exactly the ones that hold for geodesic triangles on a pseudosphere. He gave an example of one such formula, which is exactly the same, in slightly different notation, as a formula given by Lobachevsky.

In one of the great examples of noncommunication and missed opportunities in mathematical history, neither Lobachevsky nor Minding seems to have read the other's paper; *nobody* appears to have read them both and put $2i$ and $2i$ together to realize that Lobachevsky's "imaginary geometry" was nothing more nor less than the very real geometry of a particular surface. In other words, if the earth were in the shape of a giant pseudosphere, then geodesy would have led to the equations of Lobachevsky, rather than to those of spherical geometry.

What makes the lack of connection between Minding's work and Lobachevsky's all the more remarkable is that both of them refer specifically to the analogy with spherical geometry by simply using imaginary numbers in place of real ones, a link that had almost been made fifty years earlier by Johann Lambert, a younger contemporary of Euler.

Lambert, who was born in Alsace in 1728, became the

leading German mathematician of his time. His name is associated with several map projections that are now in common use in atlases around the world. Within mathematics, he is best known for having settled a 2,000-year-old question about circles: can one choose units of measurement so that both the diameter and the circumference of a circle are whole numbers? For example, if π were exactly equal to the ratio $22/7$ (rather than only approximately), then a circle with diameter 7 would have circumference exactly equal to 22 (rather than a little under 22, as is actually the case). Lambert showed that if the diameter is *any* whole number, then the circumference cannot be a whole number too. In modern terminology, π is *irrational*—not equal to a ratio of whole numbers. In another well-known work, published in 1786, Lambert came close to being the founder of non-euclidean geometry. He noted that there were two ways of replacing the parallel postulate of euclidean geometry with an alternative; he refers to them as the second and third hypotheses. In the first alternative—his "second hypothesis"—he shows that the sum of the angles of a triangle is always greater than 180° by an amount that is directly proportional to the area of the triangle, and he notes that exactly the same is true for geodesic triangles on a sphere. He then remarks that under the other alternative—his "third hypothesis"—the sum of the angles of a triangle is *less* than 180°, again by an amount that is proportional to the area of the triangle. This observation led him to write: "From this I should almost conclude that the third hypothesis holds on some imaginary sphere."

Of course, he was exactly right; his "imaginary

sphere" is precisely the pseudosphere of Minding, and his "third hypothesis" leads precisely to Lobachevsky's geometry.

Lambert focused particularly on one of the most striking consequences of his third hypothesis: there would be no such thing as similar triangles; two triangles with the same angles must actually be congruent—the angles determine the sides! The same fact about spherical triangles had been noted hundreds of years earlier by Islamic mathematicians and astronomers. For example, an equilateral triangle in the plane—a triangle with three equal sides—has three equal angles, each 60°. Any two equilateral triangles are similar—they differ just by a scaling, or magnification. An equilateral triangle on a sphere again has three equal sides and three equal angles, but in that case the size of the angles is not the same for all equilateral triangles. For small triangles, the angles are just slightly greater than 60°, while for a large triangle whose sides are one quarter the length of the "equator" on the sphere, each angle is 90°. For each angle between 60° and 90° there is exactly one size equilateral triangle on the sphere having three equal angles of the given size. In other words, two equilateral triangles with the same angles are *congruent*—they have the same angles *and* the same length sides; we cannot scale an equilateral triangle on a sphere up or down in size, keeping the same angles, as we do in the plane.

That was the property noted by Lambert under his "second hypothesis." Under his "third hypothesis," an equilateral triangle would have three equal angles, but each would be *less* than 60°. For any angle less than 60°, there is exactly one size equilateral triangle with the given size angle. Again there are no "similar triangles" having the same angles but scaled up or down in size. In a rare instance of a mathematician revealing publicly the emotional intensity he invested in his work, Lambert cites the pros and cons of accepting this consequence of his "third hypothesis." He writes:

> There is something exquisite about this consequence, something that makes one wish that the third hypothesis were true!
>
> In spite of this advantage I would not want it so, for this would result in countless inconveniences. Trigonometric tables would be infinitely large, similarity and proportionality of figures would be entirely absent, no figure could be imagined in any but its absolute magnitude, astronomers would have a hard time, and so on.
>
> But all these are arguments dictated by love and hate, which must have no place either in geometry or in science as a whole.

Despite his disclaimer, the arguments seemed to have weighed against his believing in the third hypothesis, and Lambert ended up convincing himself that it led to a contradiction and had to be discarded.

The world would have to wait half a century for the papers of Lobachevsky and Minding to spell out the ideas put forth tentatively by Lambert, and then another quarter century before the connection between those two papers was finally made. In 1868, the Italian geometer Eugenio Beltrami showed that Lobachevsky and Minding were describing two versions of the same geometry. He also achieved the goal of putting Lobachevsky's geometry on firm footing, proving that it is just as viable as euclidean geometry. He showed that if one ever did reach a contradiction derived from Lobachevsky's axioms, a similar contradiction must lurk within euclidean geometry itself. Beltrami used a two-pronged approach. First, he made the specific connection between Lobachevsky's formulas and those that hold on any surface of constant negative curvature, such as the pseudosphere. Next, he found a way to construct a "map" of Lobachevsky's "imaginary" world. Beltrami's map was drawn inside a circle in the ordinary euclidean plane; the chords of that circle correspond exactly to the straight lines of Lobachevsky's geometry. It is immediately clear that Euclid's parallel postulate does not hold, but Lobachevsky's substitute does: given any "line" and any point not on it, there are an infinite number of "lines" through the point that do not intersect the original "line." It is important to remember that Beltrami's pic-

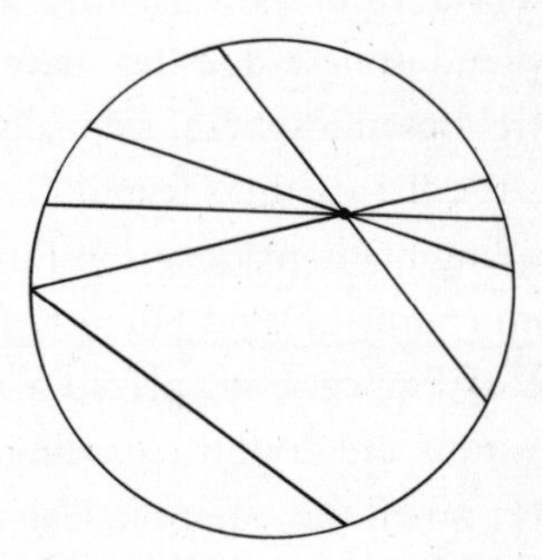

Straight lines in the Beltrami model of Lobachevsky's geometry.

ture is a *map,* not a *replica,* and just as with maps of a sphere, there are distortions of the original. On Mercator's map, infinite vertical lines correspond to finite semicircles: the meridians. On Beltrami's map, finite chords in a circle correspond to infinitely long straight lines in the original. In both cases, in order to use the map correctly, one must have precise equations that explain how measurements on the map can be translated to measurements on the original.

Strictly speaking, Beltrami's picture should be called a "model" rather than a map since the word "map" is used when there is already a surface in place that is being mapped. Lobachevsky wrote down the rules that hold in his geometry, but was not able to provide an actual playing field on which those rules applied. Some of the rules, such as the trigonometric formulas for triangles, were shown by Minding and Beltrami to hold for geodesic triangles on a surface of constant negative curvature such as the pseudosphere. However, other properties of Lobachevsky's geometry do not hold, such as the requirement that one can extend a line as far as one wishes in either direction. The pseudosphere does not have that property; if the world were a pseudosphere, we really *would* run off the edge, just as the flat-earthers feared.

The two leading mathematicians in Europe at the end of the nineteenth century, David Hilbert of Germany and Henri Poincaré of France, both played a role in settling the last few questions and doubts regarding Lobachevsky's geometry. Hilbert posed the question: Can one find a surface which, like the pseudosphere, has constant negative curvature, but, unlike the pseudosphere, does not have an edge, so that it satisfies *all* the properties of Lobachevsky's geometry? If so, it would provide the cleanest answer yet to the question of viability, without resorting to models in the manner of Beltrami. Hilbert proved that there was no such surface. One therefore has to look elsewhere for models of Lobachevsky's geometry.

Poincaré produced a new model with several advantages over Beltrami's original model. Poincaré's model shares with Mercator's map of the sphere the key property that angles in the model are correctly depicted, which is not the case in Beltrami's model. Poincaré's model can be pictured in the following manner. Cut an ordinary euclidean plane in half, say along a horizontal line. Picture the half-plane made out of some transparent material, such as glass or plastic, but with the density of the material growing progressively the nearer one gets to the boundary line. The "lines" in the geometry are the paths taken by light rays within the material. By the basic laws of refraction, those rays will be curved, and if the density is made to vary in just the right way, the light rays will follow semicircular paths perpendicular to the boundary line. Poincaré showed how to interpret measurements on this "map" to obtain an exact model of Lobachevsky's geometry.

Poincaré also produced a variant of his model, which, like Beltrami's, is entirely contained inside a circle. It is the model that has been made famous through the etchings of M. C. Escher. In fact, Escher's imaginative ways of carpeting the Lobachevsky plane provide some of the clearest and easiest means to decode Poincaré's map of Lobachevsky's geometry. In the familiar picture known as "Heaven and Hell" or "Angels and Devils" (officially entitled "Circle Limit IV"), all the angels are exact copies of each other when taken in the true scale of the Lobachevsky plane. That they appear to get smaller and smaller as they retreat

M. C. Escher's "Angels and Devils"
(officially entitled "Circle Limit IV" © 1994 by M. C. Escher/Cordon Art—Baarn—Holland. All rights reserved).

from the center of the circle is an artifact of Poincaré's map, which, in the same manner as Beltrami's, represents an infinitely long line in the true geometry by a curve of finite length on the map. What is true for the angels is also true for the devils: they are all exactly the same size and shape in Lobachevsky's "imaginary geometry," and only vary in size on the picture because Poincaré's map, like any map of the Lobachevsky plane, necessarily distorts lengths and distances.

After the work of Beltrami and Poincaré, there was no longer any question about the viability of Lobachevsky's geometry. It was transformed into one more tool in a geometer's kit, and gradually became known as *hyperbolic geometry* in contrast to both ordinary euclidean geometry and to another species of non-euclidean geometry called *elliptic geometry,* closely related to geometry on a sphere.

Hyperbolic geometry has proved enormously useful in widely divergent parts of mathematics. The most surprising may be its relation to the proposed solution in 1993 of the 350-year-old problem known as "Fermat's last theorem."

The question remains: What is the value of non-euclidean geometry in the "real world"? The answer arrived in the wake of a series of developments that began with a fundamental rethinking of the entire subject of geometry by one of the great visionaries of mathematics, Bernhard Riemann.

CHAPTER V
Curved Space

THERE IS AN ASTONISHING IMAGINATION
IN THE MATHEMATICS OF NATURE;
AND ARCHIMEDES HAD AT LEAST AS MUCH
IMAGINATION AS HOMER.

—Voltaire

Georg Friedrich Bernhard Riemann (pronounced *ree-mahn)* was born in 1826, just short of fifty years after Gauss. His birthplace, the village of Breselenz, lies in the same kingdom of Hannover that Gauss was engaged in surveying at that time. Together with Euler and Gauss, Riemann completes the trio of mathematicians who personify a golden age of mathematics in much the way that Bach, Beethoven, and Brahms are often thought to represent the pinnacle of classical music. Despite obvious differences between their careers in mathematics and music, there are some striking similarities across the two cultures.

Euler and Bach lived in the eighteenth century and, as was traditional at the time, they worked under the patronage of the nobility or royalty. They were both prolific in their personal as well as their professional lives, producing large families and voluminous quantities of manuscripts for posterity. By a curious coincidence, they also both suffered from failing eyesight in later life and ended their days in total darkness.

Georg Friedrich Bernhard Riemann
(Mathematics Institute, Göttingen).

Beethoven and Gauss, by contrast, personified the romantic ideals of the early nineteenth century. They strove to make whatever they touched a masterpiece. Bach and Euler, whose lives and dispositions seemed so much sunnier and less complicated, would sit down and compose a work in its final form at one sitting; Beethoven was famous for rewriting and reworking a piece at length before releasing it. Gauss's motto, *pauca sed matura*—"few but ripe"—reflected his obsession with searching out the essence of each idea, holding back a manuscript from publication until it fulfilled his own exacting standards. One consequence was the unhappy experience of a number of his contemporaries, who, like Bolyai, worked for years on a problem only to have the fruits of their labor elicit from Gauss a laconic response on the order of "I knew that."

Brahms and Riemann are much more solidly situated in the mid- to late-nineteenth-century world. Peering at us over their full beards, they seem far removed from the heroic hopes and pretensions of the Napoleonic era. Both were modest by nature, but found themselves in the unenviable position of being cast by their contemporaries as the natural successors to the earlier towering figures. Riemann not only was named to the professorship in Göttingen held earlier by Gauss but even took over the same lodgings in the Observatory. On the other hand (or perhaps for that very reason), Brahms and Riemann carried their drive toward perfectionism even further than Beethoven and Gauss. Brahms spent twenty years writing as many quartets before he could bring himself to publish one. His lifetime total of four symphonies contrasts with the nine of Beethoven and the one hundred of Haydn.

Riemann's few publications are even more striking, especially when compared with the outpourings of Euler. His lifetime contributions to mathematics are collected in one small paperback volume that takes up less than an inch of shelf space.

And yet Riemann was a pivotal figure whose insights

and inventions altered not only the course of mathematics but our whole worldview. Gauss's reluctance to praise Bolyai was for him much more the norm than the exception. But when he was presented with Riemann's doctoral dissertation, he embraced it immediately, speaking of Riemann's "creative, active, truly mathematical mind, and gloriously fertile originality." Years later, after Riemann's ideas had ripened and come to fruition, Albert Einstein was to write: "Only the genius of Riemann, solitary and uncomprehended, had already won its way by the middle of the last century to a new conception of space, in which space was deprived of its rigidity, and in which its power to take part in physical events was recognized as possible."

The essence of Riemann's new conception was that we should probe the space around us in exactly the same fashion that Gauss prescribed for studying a surface, by following direct paths, making measurements, and recording what we find, free from any preconceptions. We carry out what Einstein would later call a "thought experiment"—an experiment devised and carried out in our imagination.

Before describing this particular "experiment," we should say a few words about the nature of thought experiments and their important role in understanding the physical world. A good example of the role of thought experiments is the "law of inertia," first formulated by Galileo and later adopted by Newton as the first of the three famous "Newton's laws" of physics. Galileo had carefully observed and measured the motion of objects under various conditions, and finally concluded that the correct description was the exact opposite of the standard dogma that had been accepted for almost two thousand years. That dogma, promulgated by Aristotle, stated that a force was needed to maintain motion, and when the force was removed, the motion stopped. Galileo asserted that the motion would go on forever *unless* a force was exerted to stop it. The Aristotelian belief was able to hold sway for so long for the simple reason that no actual experiment could be devised to verify Galileo's assertion. There are *always* forces acting on an object: gravity, friction, and the force exerted by the earth on a falling object at the moment it hits, to name a few. Galileo had to *imagine* a situation in which all those forces were removed, and he concluded that under those circumstances an object would continue moving in the same direction at the same speed indefinitely. He was led to that conclusion by observing the effect of gradually reducing extraneous forces and seeing the actual result approximate more and more closely the prediction of his ideal experiment. The importance of this thought experiment cannot be overstated; it allowed Newton to put forces back into the picture in his second and third laws and to state the exact effect they would have on the motion of an object. As a result, the qualitative description of the physical world given by Aristotle was replaced by Newton's precise quantitative statements in the form of simple mathematical equations that became the basis of all of modern physics.

What follows, then, is not exactly Riemann's original formulation, but an equivalent thought experiment, illustrating Riemann's prescription for analyzing the shape of space. Suppose, for example, we wish to examine space in the vicinity of the earth. We first choose a direction—let us say the vertical direction at the North Pole—and investigate the "shape of space" in the plane perpendicular to that direction.

Picture a large number of rockets lined up at equally spaced intervals, along the entire length of the equator. At a given signal we simultaneously launch all the rockets. Each rocket is programmed to travel a given distance straight up before releasing a bright flare. If we had enough rockets, the result should be a giant halo of light encircling the planet at a given distance from the earth's surface. To carry out our thought experiment, let us suppose that we are able to measure the circumference of that circle. If space were truly euclidean, the length of the circle would be 2π times its radius. But Riemann's view was that we have no way of knowing in advance what the actual value of the circumference will be since we do not know if space really is euclidean. In fact, that point of view had actually been expressed earlier by both Gauss and

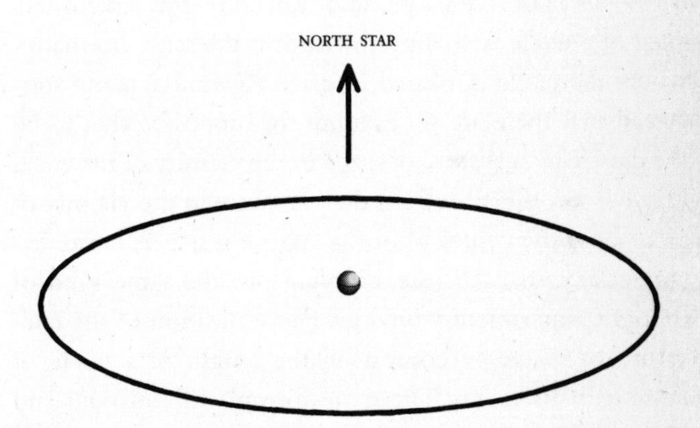

The halo encircling the earth's equator in our thought experiment.

Lobachevsky. In the 1820's Lobachevsky worked out all the formulas for non-euclidean solid geometry as well as plane geometry, including a formula that would give us the circumference of a circle of a given radius, such as the one created by the light flares in our thought experiment. In Lobachevsky's geometry (also called *hyperbolic geometry),* the circumference would be *greater* than the euclidean value of 2π times the radius, by a specific amount which would be a measure of the "curvature of space." Riemann, however, pointed out that there is no reason to presuppose that space is either euclidean or Lobachevskian. Those are but two options. It may well turn out, for example, that the length of the circle of light is *less* than the euclidean length. Riemann gave an example in which that was the case. The corresponding geometry is a different non-euclidean geometry, called variously "elliptic," "spherical," or "Riemann's non-euclidean geometry."

But Riemann went much further. In all three geometries—his, Lobachevsky's, and Euclid's—the circumference of a circle with a given radius is the same no matter where the circle is located in space. Again, Riemann suggested that there are no grounds for supposing that to be the case. The curvature of space in the vicinity of the earth may be very different from the curvature in the vicinity of a star near the center of our galaxy, or a star in some distant galaxy. In each case, carrying out the same kind of thought experiment would give an indication of the curvature of space, by comparing the length of a circle of light with its radius. The same formula of Bertrand and Puiseux that relates the length of circles on a surface to the curvature of a surface can be used to *define* the curvature of space; zero curvature means the length of a circle of radius r is $2\pi r$, positive curvature results in a length *less* than $2\pi r$—the greater the curvature, the smaller the length—and negative curvature corresponds to a length *greater* than $2\pi r$—the more negative the curvature, the greater the length.

An alternative approach to surveying space, closer in spirit to the usual process of surveying land, could be described by a somewhat different thought experiment in which the "circles of light" are replaced by a "triangulation." We could launch six probes at equally spaced points

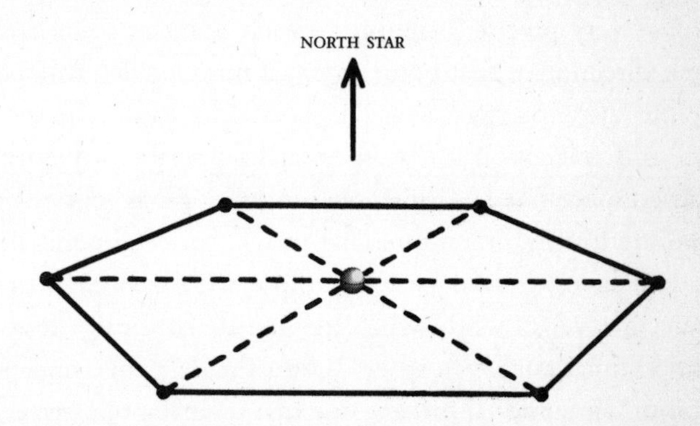

Measuring the distance between six probes, compared with the distance of each probe from earth, in order to determine the curvature of space.

along the equator, and have each of them continually monitor the distances to the two adjacent probes. If space is euclidean, then those distances at any point in its journey would be equal to the distance from the probe to the center of the earth. In Lobachevskian space the distances between the probes would grow faster than the distance the probe traveled from the earth, while in positively curved space the distances between the probes would grow less quickly than the distance covered by the probes since leaving the earth.

As with Gauss's description of curvature in terms of geodesic triangles (or the spacing of trees in an orchard), Riemann's notion of the curvature of space should be understood as a description of how measurements deviate from those that would be found if space were euclidean.

There are two common misconceptions about the curvature of space. The first is that curvature is a rather vague or qualitative concept to contrast with flatness. It is in fact very precise, assigning to each point in space and each direction at that point an exact number, determined by the shape of space near the specific location. The second misconception is that in order to describe or picture curved space one has somehow to think of it as "curving" into the fourth dimension. That image can be a useful aid in visualizing curved space for those who have familiarized themselves with what mathematicians call "four-dimensional euclidean space." Unfortunately, the concept of four-dimensional space is one that science popularizers and science fiction writers have often laced with mystical overtones. For someone who has not worked through the mathematical details, it is more likely to cloud than to clarify one's understanding. Once again the point is simply that when measurements are made in ordinary three-dimensional space, they may well not jibe with the results that are embodied in euclidean geometry; what "curvature" measures is the degree and kind of deviation from the euclidean model.

Our experience confirms that euclidean geometry offers a good description of space on a small scale. But we have absolutely no reason to assume that it also holds on a larger intergalactic scale. It is that extrapolation, from small to large scale, that makes us think in terms of a flat universe exactly as an earlier age believed in a flat earth.

Our attempts to create an accurate map of the earth offer a good analogy to the dilemma of "mapping" the universe. Maps of towns or local regions may be seemingly accurate on a small scale, but become wildly distorted on a larger scale, because no flat map can accurately depict the curve of a sphere. The same may be true of the universe. While we have learned to overcome our flat-earth mentality in measurements on a global scale, we have not yet overcome our tendency to think in terms of a flat universe.

Riemann not only invented the idea of curved space and explained how to actually compute its curvature; he also proposed a radically different model from the usual euclidean one for the entire universe. Specifically, he provided a description of the universe if it turned out to have the shape of a "spherical space." That would be the case if space had *constant positive curvature.*

We can most easily describe Riemann's universe by recalling the map of the earth depicted in the form of two separate hemispheres, each showing one "side" of the earth. Riemann's universe can be depicted in much the same manner.

Picture the earth at the center of the sphere on the left, and picture the inside of that sphere as representing all that we can currently see of the universe through our

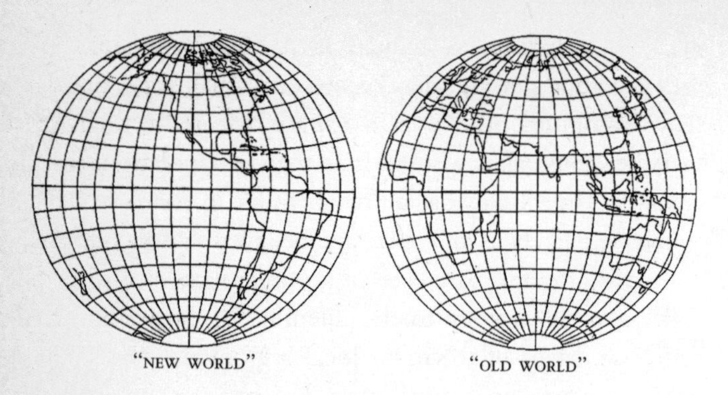

The two hemispheres of the earth.

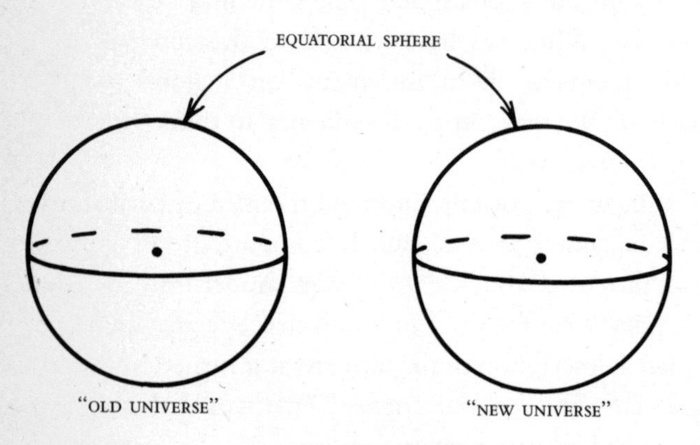

Riemann's universe.

largest telescopes. Now picture a civilization far beyond the range of those telescopes, situated at the center of the right-hand sphere, peering through their own telescopes, whose range includes everything inside the right-hand sphere.

It is easy to conceive of various possibilities: the two

spheres might be very far apart with lots of the universe between them; or they might overlap, with certain galaxies visible from both civilizations. Riemann suggests a third possibility. They might not overlap *and* they might together constitute the entire universe.

In other words, the part of the universe that we can reach with our telescopes lies inside a great sphere whose outer boundary may also be the outer boundary *from the other side* of another civilization. That outer boundary would be the "equatorial sphere" dividing the universe into two parts: the "Old Universe" that we know and the "New Universe" that a twenty-first-century space Columbus may set out to explore.

One reason that such a depiction of the universe seems to be artificial, if not impossible, is the same reason that attempts to draw accurate maps of the earth are doomed to failure: because of the curvature of the earth, the maps we draw are distortions of reality. In our pictures of the two hemispheres, we can draw our maps so that distances from the center are accurate, but then lengths of concentric circles will become more and more distorted away from the center, since the (positive) curvature of the earth's surface results in circles that are smaller than would be indicated by the corresponding circles on our flat map. On a flat map, those circles keep getting bigger and bigger, while in reality, the true circles on earth get bigger until they reach a maximum length—a great circle (corresponding to the boundary circle on the map)—and then start contracting back down toward the antipodal point.

If space had a fixed positive curvature in Riemann's sense, then the "circles of light" in our thought experiment would behave in exactly the same fashion. They would initially get longer and longer, although *less quickly* than in flat euclidean space, and eventually they would reach a maximum circumference at the outer shell of our part of the universe. They would then start to get smaller, and eventually contract down to a point at the "opposite end" of the universe, which on our map is the center of the right-hand sphere—the "antipodal point" to us in the universe. That would be the point of the universe farthest away from us. If we traveled on a spaceship, continuing "straight ahead" in any direction, we would eventually reach the antipodal point. If we kept on going past that point we would end up back where we started.

One feature of this model of the universe that particularly pleased Riemann is that it solved the age-old problem of the "edge" of the universe. Some philosophers had speculated that the universe was infinite in extent, going on forever in all directions. That theory was considered and rejected as implausible by many of those who pondered such questions seriously, from Plato and Aristotle to Newton and Leibniz. But the alternative seemed equally dubious: if it did not go on forever, then—like the flat earth—it would end somewhere; and what was beyond that?

Riemann's model resolved that paradox, which is rooted in the assumption that the universe is flat, or euclidean. If instead it is positively curved and Riemannian, then it can be finite in extent and still not have any "edge" or "boundary." In Riemann's model, every part of the universe looks just like every other part, as far as shapes and measurements go.

At the time that Riemann first presented this picture as a possible description of the real world, it must have seemed little more than a creation of his "gloriously fertile" imagination. And even today—a century and a half later—it stretches our imaginations to the limit to encompass Riemann's vision.

There is a much quoted story about David Hilbert, who one day noticed that a certain student had stopped attending class. When told that the student had decided to drop mathematics to become a poet, Hilbert replied, "Good—he did not have enough imagination to be a mathematician."

In a rare confluence of the poetic and mathematical imagination, the poet Dante arrived at a view of the universe with striking similarities to that of Riemann. In the *Divine Comedy,* Dante describes the universe as consisting of two parts. One part has its center at the earth, surrounded by larger and larger spheres on which move the moon, the sun, successive planets, and the fixed stars. The outer sphere, bounding all the visible universe, is called the *Primum Mobile.* What lies beyond is the "Empyrean," which Dante pictures as another sphere, with various orders of angels circling on concentric spheres about a center where a point of light radiates with almost blinding intensity.

The poet is led by Beatrice from the surface of the earth, through the various spheres of the visible universe, and all the way to the *Primum Mobile.* Looking out from there, he finds himself looking *in* to the sphere of the Empyrean. There is no indication that one must choose a particular point on the *Primum Mobile;* presumably, looking out at any point would give a view into the Empyrean. In other words, we are to think of the Empyrean as somehow both surrounding the visible universe and adjacent to it. If that is the case, then the universe according to Dante would coincide exactly with the universe according to Riemann; they would differ only in the labels.

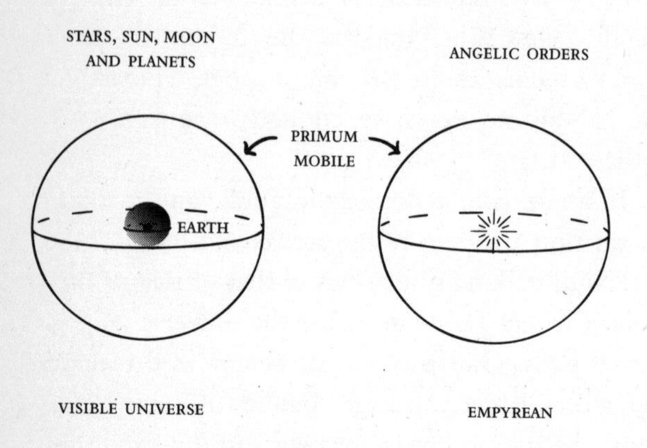

Dante's universe.

Riemann's vision is of course more "scientific" than Dante's, in that it is quantitative as well as qualitative—Riemann gives formulas from which one can derive the area of concentric spheres, the circumference of circles, and so on.

The shape of the Dante-Riemann universe is what mathematicians call *spherical space* or a *hypersphere.* It is like the ordinary sphere—elevated to a higher dimension. The analogies are clear; concentric circles on an ordinary sphere initially get larger, reach a maximum size and then start to get smaller. On a hypersphere, concentric (ordinary) *spheres* start by getting larger, reach a maximum size, and then contract down. On both the sphere and the hypersphere, starting at any point in any direction and continuing "straight ahead" will eventually lead back to the starting point. Furthermore, the total distance traveled will be the same, no matter what the starting point and direction.

A sphere and a hypersphere can be of any size; the size is determined by the total length of a round trip from any point. The length of the round trip also determines the curvature: the longer the round trip, the smaller the curvature and the closer the geometry is to euclidean.

Riemann's conception of a spherical space, together with his suggestion that such a space may describe the actual shape of our universe, constitutes one of the most original and radical departures from the standard worldview in the history of science. A leading twentieth-century physicist, Max Born, has said: "This suggestion of a finite, but unbounded space is one of the greatest ideas about the nature of the world which ever has been conceived." Ironically, Born thought he was referring to an idea of Einstein, since Einstein incorporated Riemann's spherical space into his work on cosmology, along with two other fundamental ideas of Riemann: the curvature of space and the description of a curved space of four dimensions. Riemann invented all these concepts as well as an important alternative to spherical space: "hyperbolic space"—of equal interest to modern cosmologists—while still in his twenties. He presented them at Göttingen in 1854, at the age of twenty-eight—in a lecture that we now see clearly in retrospect as marking the birth of modern cosmology.

As it turned out, Riemann was still missing one key element for a complete picture of the universe. For that the world would have to wait another half century, for the birth of Albert Einstein.

The route from Riemann to Einstein was anything but direct. The first steps were taken by Riemann himself. While still in his early twenties, Riemann set himself the task of developing a unified mathematical theory connecting electricity, magnetism, light, and gravitation. The very notion that there might be a "unified field theory" was so far ahead of its time that even a century later Einstein would be mocked for spending his latter years in a vain search for it.

We will never know along what lines Riemann's ideas might have evolved had he lived a normal life span. Tragically, he died before his fortieth birthday, like Mozart and Schubert before him. In the last year of his life a paper was published that was destined to revolutionize our understanding of the physical phenomena that had resisted Riemann's best efforts, that would realize a key part of his vision of unification, and that would prove an important link between Riemann and Einstein. The dramatic manner in which these discoveries were made provides the next episode of our story.

CHAPTER VI

The Invisible Universe

WHAT MEN ARE POETS WHO CAN SPEAK OF JUPITER IF HE WERE LIKE A MAN, BUT IF HE IS AN IMMENSE SPINNING SPHERE OF METHANE AND AMMONIA MUST BE SILENT?

—Richard Feynman

Unlike the divisions of time into days and years, which correspond to natural phenomena, such as the rotation of the earth on its axis and the orbit of the earth around the sun, the concept of a century has no physical basis. Rather, it derives from an arithmetical artifact built upon an anatomical accident. If human beings came equipped with four fingers on each hand, then we would undoubtedly have developed a number system with base eight rather than base ten, leading to divisions of time into octads rather than decades and intervals of eight octads—or sixty-four years—instead of a century. The whole ten-fingers-inspired decimal system and the resulting grouping of years into decades, centuries, and millennia are so convenient as an organizing principle to which we attach convenient labels—the Gay Nineties, the Roaring Twenties, the turbulent sixties—that we lose track of the arbitrariness of the divisions. (The true "sixties" culture in the

States, to the extent that there was one, more erly embraced the octad from 1964 to 1972.) Simi- rly, individuals dread turning forty (or thirty, or fifty), investing a particular number with unwarranted significance just because it is a multiple of ten.

Despite that disclaimer, nineteenth-century science had a distinctive character and played a key role in our attempts to "see" and understand the universe. Two pivotal events took place in the year 1800 that set the agenda for much of science during the remainder of the century. The first was the discovery of infrared radiation by the eminent astronomer Sir William Herschel, followed shortly, in 1801, by the discovery of ultraviolet rays by Johann Wilhelm Ritter. As a result, scientists gained a completely new perspective on the nature of light. They realized that visible light was bracketed by other types of rays that were equally real, but had previously eluded our awareness, because they were invisible.

The second major scientific event of 1800 was the invention by Alessandro Volta of the "voltaic pile"—the forerunner of all electric batteries. For the first time scientists had available a reliable source of electricity with which to carry out basic experiments. Soon electric currents were used to decompose water into its constituent elements, which turned out unexpectedly to consist of a pair of gases—hydrogen and oxygen. But progress was agonizingly slow and understanding the nature and properties of electricity was a long-drawn-out process. Not till near the end of the century did such practical applications as electric lights become a reality. Also toward the end of the century, further discoveries were made that set the stage for twentieth-century science and technology. In 1885, Heinrich Hertz used electrical discharges to produce a new kind of radiation, later designated "radio waves." Surprising applications followed, in the form of long-distance "wireless" communications. The strangest and most unexpected discovery was undoubtedly that of X-rays in 1895 by Wilhelm Roentgen. He was experimenting with electric currents traveling through a vacuum in a glass tube when he realized that some form of radiation was escaping not only through the glass but through normally opaque materials. Victorians were shocked and amazed to see pictures taken right through clothes revealing shadowy internal organs and sharply etched bones. Roentgen and others quickly realized the value of X-rays in medicine, and by 1900 their use was widespread. Roentgen's discoveries earned him the first Nobel Prize in physics in 1901.

Even more surprising than the discoveries themselves may be the fact that both X-rays and radio waves were "discovered" mathematically long before they were observed as physical realities. In fact, it was precisely because radio waves had been predicted in advance that Hertz set out to confirm their existence experimentally.

The credit for these revolutionary discoveries goes to James Clerk Maxwell, who at mid-century was actively investigating all of the principal research areas of physics: gases, fluids, electricity, magnetism, optics, etc. He combined a powerful physical intuition with an extraordinary mathematical ability to produce definitive equations describing a variety of physical phenomena. His most fa-

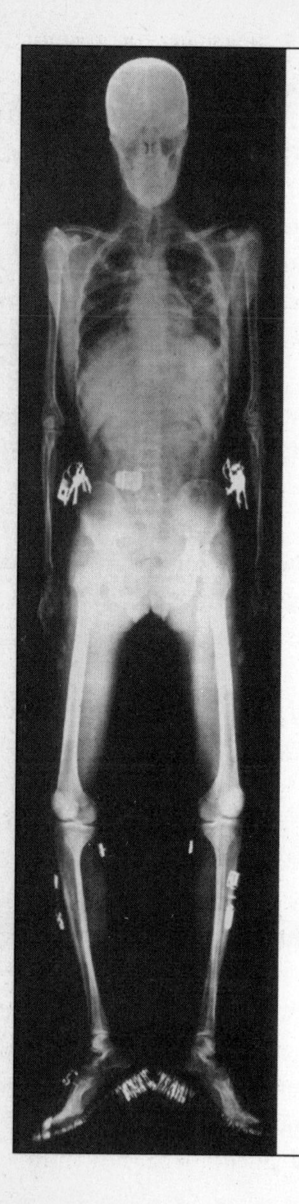

The first full-body X-ray picture of a person. (Deutsches Museum, München)

mous achievement was the integration of electricity and magnetism in a single set of equations.

Maxwell's equations seemed to be imbued with the same almost uncanny power to encompass all of electric and magnetic phenomena that Newton's equations, two

James Clerk Maxwell with his wife, Katherine Mary Dewar, a few years before his untimely death at age forty-eight. (University of Cambridge, Cavendish Laboratory)

hundred years earlier, had displayed with regard to planetary motion and mechanical actions. Maxwell's equations also had the potential to predict new and unforeseen phenomena, the most important of which was the concept of an "electromagnetic wave" consisting of electric and magnetic vibrations traveling at a high speed. Maxwell was able to calculate their speed of travel using experimentally determined quantities from both electricity and magnetism; the answer came out to be very close to the known value of the speed of light. As a result, as a by-product of his search for a unified theory of electricity and magnetism, Maxwell inadvertently achieved a part of Riemann's plan to integrate the theory of light with the other two phenomena: light, he discovered, was itself a form of electromagnetic radiation. In addition, Maxwell concluded that there should be other electromagnetic waves, all traveling at the same speed, but vibrating at different frequencies. Twenty years later his prediction was confirmed experimentally by Heinrich Hertz, who lauded the almost magical ability of Maxwell's equations to predict new physical phenomena: "One cannot escape the feeling that these mathematical formulas have an independent existence and an intelligence of their own, that they are wiser than we are, wiser even than their discoverers, that we get more out of them than was originally put into them." Maxwell's own evaluation of his work was expressed more succinctly in a letter written early in 1865: "I have also a paper afloat, containing an electromagnetic theory of light, which, till I am convinced to the contrary, I hold to be great guns."

Neither Maxwell nor Hertz nor anyone else at the

time would have dreamed that out of those few equations would grow the future industries of radio, television, and radar, as well as countless other scientific and technological applications.

X-rays too, it was realized, were a kind of electromagnetic wave. Gradually scientists learned how to produce other forms of electromagnetic waves as well, such as microwaves and gamma rays. But perhaps most surprising of all was the discovery in the twentieth century that all these forms of electromagnetic waves had in fact been surrounding and bombarding us throughout history, penetrating the earth's atmosphere from outer space. By a quirk of physiology, only a tiny fraction of those waves are directly perceived by us as visible light; all the others require special instruments for us to detect and to convert into a form that we can either see or hear.

Fifty years ago Grote Reber inaugurated a new era for astronomy with the publication of a paper entitled "Cosmic Static." Reber had set up a homemade radio telescope in his backyard in Wheaton, Illinois, and had produced a contour map of the part of the sky visible from that position, based on the strength of the "cosmic static"—that is, the radio waves he received from outer space.

The story of astronomy in the second half of the twentieth century is closely bound up with the development of new telescopes and instruments for "seeing" infrared, ultraviolet, X-rays, microwaves, and all the other forms of radiation, giving us an entirely new view of the universe. Objects long familiar in the visible-light range take on entirely new aspects when viewed with infrared and other telescopes. An astronomer today would feel blindfolded if restricted to observing a galaxy, quasar, black hole, or other deep-space object in the visible-light range. Dramatic discoveries have been made of radio and X-ray sources in the sky that do not correspond to any known object in the visible spectrum. A comprehensive picture of any celestial object requires images from as wide a representation of wavelengths across the spectrum as possible.

Seeing and observing, however, are only the first steps toward understanding. The human race observed the Milky Way for millennia before realizing that what we were seeing was the edge-on view of a gigantic rotating swirl consisting of billions of stars, each comparable in brilliance to the sun. In order to arrive at a comprehensive view of the universe, we still have to provide the geometric framework that will allow us to integrate and interpret the wealth of observations made possible by nineteenth-century discoveries and twentieth-century technology.

CHAPTER VII

Looking Back: The Observable Universe

THE WORLD OF IDEAS WHICH IT DISCLOSES OR ILLUMINATES, THE CONTEMPLATION OF DIVINE BEAUTY AND ORDER WHICH IT INDUCES, THE HARMONIOUS CONNEXION OF ITS PARTS, THE INFINITE HIERARCHY AND ABSOLUTE EVIDENCE OF THE TRUTHS WITH WHICH IT IS CONCERNED, THESE, AND SUCH LIKE, ARE THE SUREST GROUNDS OF THE TITLE OF MATHEMATICS TO HUMAN REGARD, AND WOULD REMAIN UNIMPEACHED AND UNIMPAIRED WERE THE PLAN OF THE UNIVERSE UNROLLED LIKE A MAP AT OUR FEET, AND THE MIND OF MAN QUALIFIED TO TAKE IN THE WHOLE SCHEME OF CREATION AT A GLANCE.

—J. J. Sylvester,
nineteenth-century mathematician

To Dante, as well as to early civilizations, and, for that matter, to modern civilizations until relatively recent times, the stars were scattered on the surface of the heavenly sphere like so many jewels encrusted on the ceiling of a temple. Three profound insights revolutionized the way we interpret what we see as we gaze upward on a starry night.

The first realization was that the stars in constellations such as the Big Dipper are not just laid out on a surface, but are located at vastly different distances from the earth. The figures that we form in our mind's eye appear like the skyline of a city in the distance, as if painted on a backdrop, but in reality the stars are distributed in three-dimensional space. They would look very different to an observer in space or on some other star in our galaxy. The Gemini "twins," Castor and Pollux, for example, appear from the earth to be "twins"—almost identical in brightness—but in fact Castor is much farther away than Pollux, and appears equally bright because it is twice as luminous.

The notion that stars might be located at different distances from the earth was an old one, but the actual distances involved turned out to be so vast that until astronomical instruments attained a certain level of precision, there was no way to verify, much less measure, the variation in distances of the stars. It was not until 1838 that the astronomer and mathematician Friedrich Wilhelm Bessel was able to detect the apparent displacement of a star relatively close to the earth against the background of more distant stars, as the earth, over a six-month period, moved from one side of its orbit around the sun to the other. Using the width of the earth's orbit, and a simple application of similar triangles, it was then easy to estimate the stellar distance.

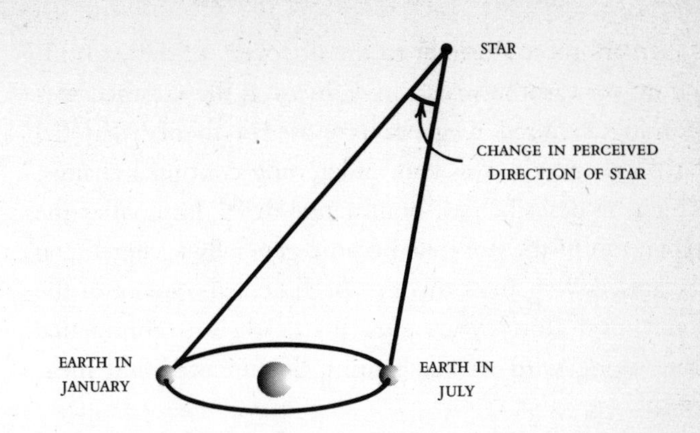

Change in the perceived direction of a star, used to calculate its distance from earth. (The change in direction on the picture is far greater than in actuality.)

The second crucial insight regarding the skies was that, because distances to stars vary, and because light travels at a finite speed, the light we see from each star originated at a different moment in the past; looking out in space and looking back in time are one and the same. The farther away a star is, the farther back in time we are seeing. The disparity between stellar and terrestrial distances is so great that the use of miles or kilometers becomes unwieldy; even "nearby" stars are located hundreds of trillions of miles away. It is more convenient to measure stellar distances in terms of the distance light travels in a year. As a result, the nearest star is described as being about four "light-years" from the earth. A light-year is roughly six trillion miles, or ten trillion kilometers.

It was not until the twentieth century that the third and most unexpected discovery was made. It turned out that the universe was not static, as had always been assumed, but is in a state of rapid expansion.

In an ironic sidelight to the discovery of the expanding universe, some years earlier, in 1912, the German meteorologist Alfred Wegener proposed a theory that the earth's geography was also undergoing continual change, which he described as "continental drift." Long after the expansion of the universe became generally accepted and acclaimed, Wegener's theory was ridiculed. Not until the 1960s, over thirty years after his death, was continental drift, along with its mechanism, documented and measured.

The discovery of the expanding universe is often attributed to Edwin Hubble, who in 1929 presented the observational evidence that distant galaxies appear to be receding at speeds proportional to their distance. However, the true story of that discovery is quite different, and far more fascinating. It is the story of a wonderful interplay between theory and observation taking place over a dozen years from 1917 to 1929. The first hints that the universe was expanding were contained in theoretical papers on cosmology published in 1917 by Albert Einstein and the Dutch astronomer Willem de Sitter. Both papers were based on Einstein's general theory of relativity, formulated two years earlier in 1915. Einstein had written down a set of equations describing the workings of gravity, and when he tried to apply those equations to the study of the entire universe, he discovered that they appeared to be incompatible with a static universe. Since there was no evidence then that the universe was not static, Einstein added an extra term to his equations that allowed him to find a solution of the equations in the form of a model of the universe that was unchanging with time. De Sitter found a solution of Einstein's equations of a different nature; one aspect of his solution was that distant galaxies should appear to be receding at a rate that increased with their distance. (Hubble acknowledged the key role of de Sitter's theoretical work in a letter he wrote to de Sitter in 1930. "The possibility of a velocity-distance relation among nebulae has been in the air for years—you, I believe, were the first to mention it.")

Between 1917 and 1929, a series of theoretical and observational advances shifted opinion more and more in favor of an expanding universe. One key player was Vesto M. Slipher, who in 1912 had already started on a program to measure the velocities of galaxies. By 1922 he had accumulated a list of forty-one galaxies; except for a handful of nearby galaxies, they all appeared to be moving away from our own galaxy.

In the same year, a brilliant Russian scientist, Alexander Friedmann, found a solution of Einstein's original equations of relativity—without the additional term that Einstein had introduced to keep the universe from expanding—in which the universe goes through phases of expansion and contraction. The following year, Hermann Weyl—the leading German mathematician of his day—devised a new approach to cosmology using relativity theory that led him directly to a relation implying that galaxies recede at velocities that increase with their distance.

The missing link in all these discussions was a reliable way of measuring distance to far-off galaxies. It was to this problem that Hubble devoted a large part of his energies during the 1920s. Working with the world's most powerful instrument at the time, the 100-inch telescope on Mount Wilson in California, he first made a major discovery in 1923, when he identified a particularly important kind of star, known as a Cepheid variable, in the Andromeda galaxy. Cepheid variables were at that time one of the most reliable indicators of astronomical distances. Hubble's 1923 discovery served simultaneously as the first step toward settling the question of the expanding universe and as the final clue in settling the even more fundamental question: What are the building blocks of the universe? A great debate had raged during much of the previous decade over the size of our galaxy and whether it encompassed the entire observable universe. The focal point of the debate was the fuzzy spots of light known as "nebulae" that had been discovered in increasing numbers as telescopes became more and more powerful. Both the size of our galaxy and the distances of these nebulae were exceedingly difficult to determine, leaving open the question whether all the nebulae were located within our galaxy or whether some of them were at great distances and constituted "island universes" of stars, or galaxies, of their own. Prior to 1918, astronomers grossly underestimated the size of our galaxy. It was Harlow Shapley who first proclaimed, in 1918, that our galaxy was some hundred times larger than previously thought; the ideas he outlined on the approximate size and shape of the galaxy have since been confirmed and universally accepted. But his very success in conquering the mysteries of our galaxy led Shapley astray on the issue of external galaxies. He considered the weight of evidence to fall on the side of those who believed that all the nebulae lay within the confines of our galaxy; in other words, that our galaxy constituted the entire universe.

And so, for a time, coming to grips with the question of whether the universe was expanding was made more difficult by our rather slippery grasp on the very nature of that universe. When Hubble was able in 1923 to place the "Andromeda nebula," as it was then called, well beyond the bounds of the Milky Way galaxy, the matter was effectively settled. It was also a momentous occasion for amateur stargazers since the famous blur of light in the constellation Andromeda became the first (and only) celestial object visible to the naked eye from the Northern Hemisphere that was shown to lie outside our home galaxy.

Between 1923 and 1929, Hubble went on to fix the distances to another twenty-three galaxies and to obtain what he considered to be likely estimates for an additional twenty-one galaxies. During the same period, two of the leading theoretical cosmologists—Georges Lemaître in Belgium and Howard Robertson in the United States—published papers in which they showed that a universe expanding in accord with Einstein's equations should have the property that distant galaxies would be receding at velocities proportional to their distance. Once Hubble had obtained a relatively reliable set of distances to a number of galaxies, he was able to compare them with the velocities found earlier by Slipher and others and thus set the velocity-distance relation—now known as Hubble's law—on a solid footing.

It is very likely true that not since Newton formulated his law of universal gravity has any comparably simple physical law led to such an astonishing series of consequences. Hubble's law radically altered not only our understanding of the evolution of the universe but also our interpretation of what we are seeing right now. Indeed, the two are indissolubly linked, since what we see now *is* the past history of the universe. Perhaps we need a name for the entity whose elusive geometry we wish to describe. Let us call it the *retroverse.* It is the part of the whole universe that we observe looking out—which is to say, looking back—from a particular place—the earth—at a particular time. And we "observe" with all the instruments at our disposal, via visible light, ultraviolet, infrared, microwaves, radio waves, X-rays, gamma rays, the whole spectrum. The question is: What is the true shape of what we observe? What is the geometry of the retroverse?

To answer that question, we need to examine Hubble's law in more detail. It states that distant galaxies are moving away from us with a speed roughly proportional to their distance.

The statement of Hubble's law can be dissected into three parts:

1. Other galaxies are receding from us.
2. The rate at which they recede depends on their distance.
3. There is a constant ratio—the "Hubble constant"—between their velocities and their distances from us.

To see what the first two of these statements mean in practice, think of a number of galaxies at roughly the same distance from us, all situated near the horizon. According to the second statement in Hubble's law, they are all receding from us at roughly the same velocity.

The third statement implies that the galaxies situated twice as far away as the first are moving away twice as fast as the first.

Galaxies at the same distance from the earth, receding at the same velocity.

An immediate consequence is that the distance between the inner and outer rings of galaxies is also increasing, and at exactly the same rate as the distance between ourselves and the inner ring. In the same way, if we take a

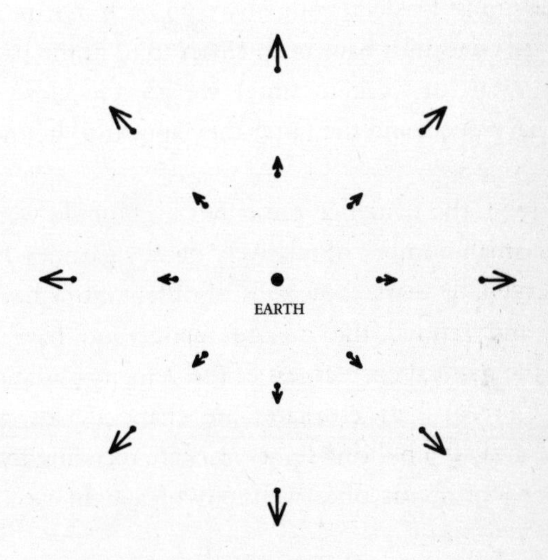

A second ring of galaxies, moving away from the earth at twice the speed of the first ring.

whole series of rings spaced at equal intervals, then each successive ring moves away faster than the previous one, so that the spaces between them are all growing at the same rate.

So the first consequence of Hubble's law is that although it seems to refer only to the relation between us and other galaxies, it also implies that those other galaxies are receding from each other in the same manner. In other words, observers on some other galaxy would arrive at the same "law" referred to their home galaxy that we have found for ours.

The most dramatic consequence of Hubble's law is what it tells us about how we got to where we are now. Simply play the Hubble tape in reverse: if distances between galaxies are increasing as we look toward the future, then they must be decreasing as we go back in time. Each ring of galaxies must have been closer to us in the past; the further away (or back in time) we go, the closer they would have been, and the faster they appear to be moving toward us.

In 1929, the evidence presented by Hubble was limited to a small number of relatively nearby galaxies. But in the intervening years, thousands of observations have extended and refined the measurements, and have confirmed the general correctness of the velocity-distance relation. Current best estimates are that galaxies whose distance away is a billion* light-years are receding from us at a speed of about one-twentieth of a light-year each

* I use a "billion" in the American sense of a "thousand million" rather than the British usage of a "million million."

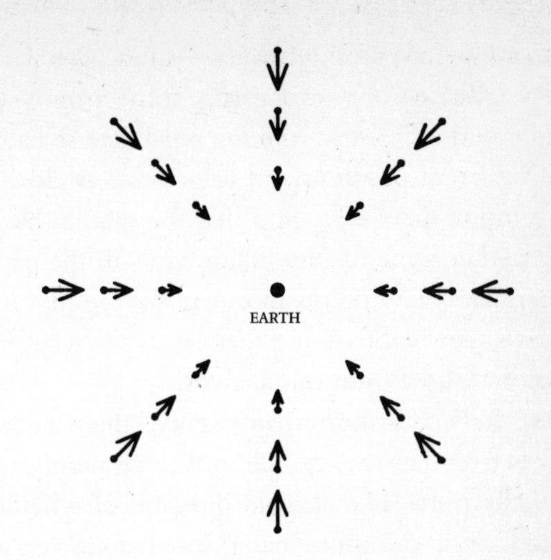

Going back in time, galaxies will come closer together.

year. Assuming for the moment that there is no reason for that speed to have changed over the course of time, those galaxies must have been a twentieth of a light-year closer to us each year in the past. To have ended up a billion light-years away, they must have started at exactly the same point as we did, some twenty billion years ago! As for galaxies that are two billion light-years away, they are receding twice as fast, or one-tenth of a light-year per year, and in the past they were that much closer each year. Starting together twenty billion years ago and moving apart at one-tenth of a light-year per year would bring them to their present distance of two billion light-years away.

In exactly the same fashion, the fact that the speed is proportional to the distance leads to the conclusion—as indisputable as it is remarkable—that every galaxy that we can observe—our entire retroverse—must have started in the same place as our own galaxy some twenty billion years ago. Said differently, running our universe backward from the current positions and velocities—as close as we can determine them—we find that the whole thing collapses together some twenty billion years in the past.

There are a few points in our reasoning that merit a closer look, but before doing that, let us see what further conclusions follow from this analysis.

First: *Nothing* is more than twenty billion light-years away. It is true that twenty billion is a big number, but it is decidedly finite. The age-old question of whether the universe is finite or infinite has a clear-cut answer: insofar as Hubble's law holds—as all current evidence indicates to be the case—the universe is finite, and we have an explicit bound on its size.

Of course, by "universe" here we mean the observable universe: the retroverse. It is not clear if it even makes sense to pose the question for those parts of the universe that may be somewhere "out there" but that we cannot observe. (We may try posing it in the next chapter anyway.)

Second: Look again at our picture of the part of the retroverse that lies along the horizon. We may start by using concentric rings of galaxies at intervals of a billion light-years.

Each ring of galaxies is an additional billion light-years from the earth.

What does the outer ring in this picture represent? Since nothing we observe can be more than twenty billion light-years away, there must be exactly twenty rings in our picture, and anything in the outer ring is at a distance of twenty billion light-years. The light or other radiation we are receiving from it was emitted twenty billion years ago. But twenty billion years ago all the observable objects were collapsed to a single point. So the outermost circle in our picture must represent a single point in the universe.

We are now faced with an apparent paradox. The circles of galaxies in the illustration seem to be growing larger and larger as they get further away from us, but the outermost circle should in fact be pinched together in a single point. However, the paradox is only apparent. It stems from an implicit assumption that the picture that we have drawn is drawn to scale. In actual fact, the picture does represent correctly distances from other galaxies to our own, which means that it is drawn to scale along each line from the center to the outer rim. Also, directions toward individual galaxies are correctly indicated. But distances and angles are accurate only in relation to the center. Distances between two faraway galaxies may be greatly distorted.

These properties are strongly reminiscent of the situation we encountered in trying to draw a map of the earth's surface. The picture we have drawn is precisely what we described in Chapter 2 as an "exponential" or "egocentric" map. Just as in the map shown there, where the outer circle represented a single point on the earth—the antipodal point to the one at the center of the map—here, too, the outer circle on our galactic map represents a single point in the universe—the point where everything we observe was concentrated twenty billion years ago. The apparent paradox is resolved if the slice of the universe that we see, looking out in all directions around the horizon, has a shape that, at least schematically, is a sphere. Since the scale of our "map" is correct along lines through the center, every point on the next-to-

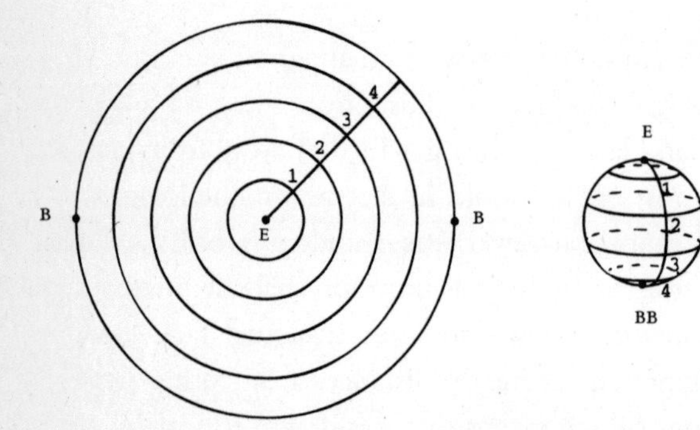

Rings of galaxies on an egocentric map (left), and the same rings portrayed on a sphere (right).

outermost circle is at a fixed distance from a point on the outermost circle. But the whole outer circle corresponds to a single point—the one that we have labeled *BB* in the drawing above—and as is apparent, the adjacent circle does correspond to all points at a fixed distance from *BB*. Similarly, the second circle, counting in from the outermost one, corresponds to all points at twice that distance from *BB*.

Yet the true shape might be more like a turnip or a pear. All of the requirements of the egocentric map are equally fulfilled in the two drawings below: We have no direct way of knowing what the precise shape is, since we can only measure distances *from us.* But the general picture is clear.

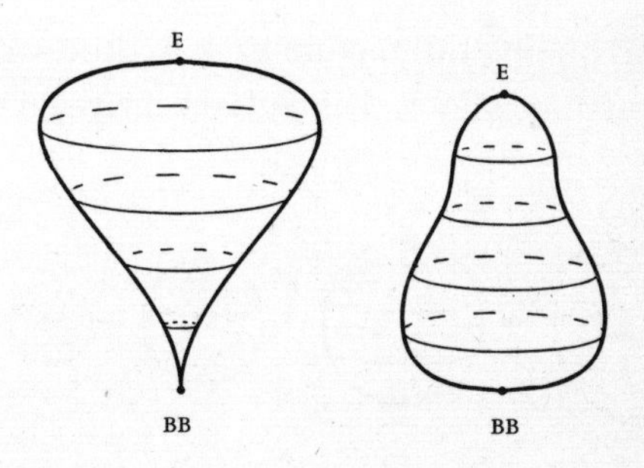

Here is another possible paradox: According to Hubble's law, points on successive circles of galaxies are moving away from our own at ever increasing rates. How can the outer circle, corresponding to the point *BB* on the sphere, be moving away both to the right and to the left?

The answer is simple. In the sphere (or pear, or turnip), the whole surface is expanding, just like a balloon.

So the distance from us (at *E)* to *BB* is increasing *in all directions.*

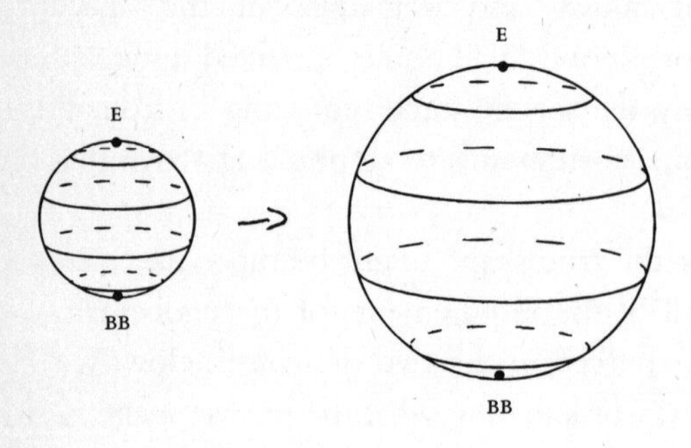

The view back, now and later.

Again, running this picture back in time, we can visualize a shrinking balloon that ends up shriveled to a point.

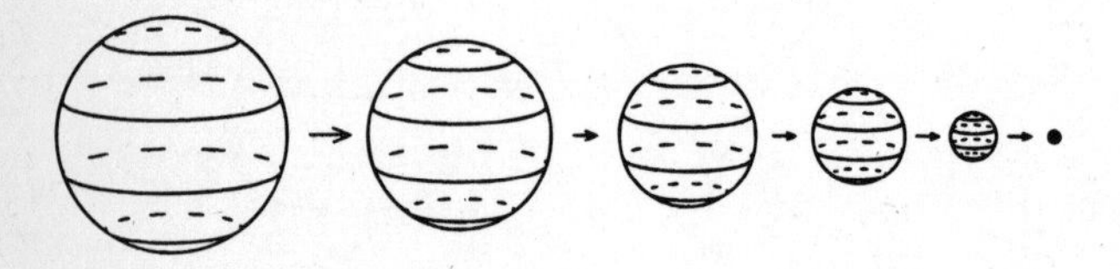

The changing view, going back in time.

The main point of this tale is that whatever the exact shape of this slice of the retroverse may be, it is *not* flat or euclidean; if it were, then the egocentric picture we drew would be an exact scale representation of reality. But on such a picture the concentric circles of galaxies become larger and larger, whereas in the actual universe they initially get larger, but eventually start to shrink, finally contracting to a point. That behavior is a sure sign of positive curvature. And so we come to the last of our initial round of consequences derived from Hubble's law: The retroverse cannot be euclidean, but must have, at least somewhere, positive curvature in the sense of Riemann.

To arrive at a picture of the entire retroverse we follow the same line of reasoning, but now look out in all directions, rather than just along the horizon. We then find that all galaxies lying on a sphere a billion light-years away are receding from us at a speed of roughly one-twentieth of a light-year per year. Those galaxies twice as far away are receding at twice the speed, and so on. It follows, as before, that if we run our imaginary clock backward, each of those spheres, as well as everything in between, will converge to a single point at a particular moment in time, approximately twenty billion years ago. What we are now seeing—the receding galaxies in all directions—can be thought of as the aftermath of a gigantic explosion of a magnitude impossible to conceive: the legendary big bang. Also impossible to conceive is that all the matter in the entire observable universe was once condensed into a sphere the size of a pinhead. Nevertheless, the evidence points to that being the case. As a result, looking back to spheres of galaxies more and more distant, we first see larger and larger spheres, but as we look further back in time, toward the big bang, the spheres must eventually contract back down. We may therefore picture the entire retroverse in two halves: the first, as the concentric spheres about the earth get larger and larger, and the second, as they contract to the big bang. Taken together, they determine a geometric figure we have met before—the hypersphere. The resemblance to Dante's picture of the universe in Chapter 5 is almost eerie; the big bang occupies the position where Dante placed a point of light radiating with great intensity.

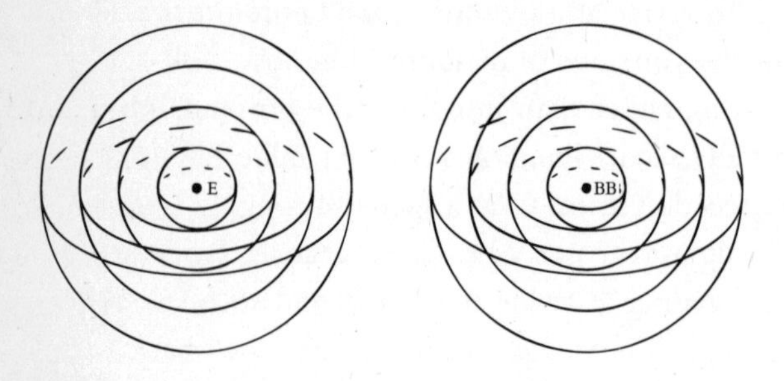

The retroverse as a hypersphere.

Before going any further, we should address the question: Have we made any unwarranted assumptions that may have led us astray and invalidate the conclusions?

The answer is both "yes" and "no." We have indeed made some unwarranted assumptions, but although they require some adjustments in details, they leave the overall picture unchanged.

The first and most fundamental assumption is that Hubble's law is correct. The very notions of "distance" and "velocity" become harder to pin down when we contemplate galaxies separated by cosmological distances of billions of light-years, each moving at an ever changing rate over the period that it takes light from one to reach the other. At the present time, the overwhelming majority of physicists are convinced of the truth of Hubble's law, but a few skeptics remain. So we must take our scenario as the most likely one based on present evidence, and allow for the possibility of new observations or interpretations leading to a major revision. However, the main conclusion we have drawn from Hubble's law—the origin of the universe in something like the big bang—has some fairly convincing physical evidence in its favor.

The second assumption we made is that, in estimating the age of the universe, the rate at which galaxies are speeding away from us was unchanged over time. That is indeed an unwarranted assumption, since we would expect the gravitational attraction between galaxies to have some braking effect, slowing down the rate of expansion of the universe. In other words, the velocities at which galaxies are receding should have been greater in the past than the ones currently observed. Taking that into account as we run the universe backward in time, we find that the broad picture of the universe is unchanged, but that there is a shorter interval back to the big bang. Current estimates are more on the order of ten or twelve to fifteen billion years, rather than twenty.

The third assumption we have made is that we have talked about Hubble's law regarding the velocities of receding galaxies as if it applied all the way back to the beginning of time. Yet that is not the case. We can directly observe galaxies that lie some 90 percent of the way back to the big bang. One of the great hopes for the new instruments being introduced during the present decade is that they will allow us to explore the distant parts of the universe beyond the range of current technology. What exactly we will find is open to speculation, but it is almost certain that we will not find many more galaxies. The reason for that is one of the great ironies of our unfolding picture of the universe. Hubble's law, derived from observing the apparent motion of distant galaxies, leads to the conclusion that there *are* no galaxies beyond a certain distance. The reason has to do with the physical consequences of the big bang scenario, rather than the purely geometric ones that we have described so far. As we run time backward and distances between galaxies contract, we find that the average temperatures must rise. And as we retreat further and further into the past, we reach temperatures where not only could galaxies not exist, but the very elements out of which the individual stars in a galaxy are made cannot be sustained; individual atoms would be broken up into their component parts—free electrons, protons, and neutrons. Thus, in the early stages of the universe, there could be no galaxies, and Hubble's law as we stated it would be meaningless. However, other ways of describing the expansion of the universe are still possible, using, for example, the change in the average distance between particles. In any case, the precise nature of the universe during the era roughly between a million and a billion years after the big bang, when galaxies first formed, is one of the principal remaining mysteries that the array of new instruments expected to be put into place may be able to help resolve by the year 2000.

The fourth assumption, almost certainly false, is that we will eventually be able to "see" essentially all the way back to the beginning of time. In fact, for the physical reasons mentioned above, it is likely that we will never receive radiation of any kind—light, radio, or whatever—from the first few hundred thousand years after the big bang. The most distant messages we will ever get are likely to emanate from the same era as those that we now have, called the "3-degree microwave background radiation," which is generally thought to have originated about three hundred thousand years after the big bang. According to current understanding, the state of the universe further back than that was somewhat like the interior of a star, where photons of light are absorbed as quickly as they are emitted, and the result is an opaque mixture of radiation and particles: a very hot primeval soup that had to cool down sufficiently to allow radiation to escape.

And so our story of the retroverse has a surprise ending. A curtain suddenly drops, permanently blocking from view what lies beyond. A picture of that curtain was first obtained in 1992—a fitting commemorative for the half millennium since Columbus arrived in the "New World." The picture (on page xiii) was a computer-drawn composite of millions of observations made by instruments specially designed for the purpose and placed on board a satellite named COBE (Cosmic Background Explorer). COBE was launched late in 1989, just twenty-five years after the microwave background radiation was first discovered. It took years of gathering and analyzing data before the final picture could be produced.

That picture was reproduced on front pages of newspapers around the world, displacing news that was merely local, national, international, or galactic. It was a momentous occasion, when earthlings first saw a picture obtained by looking back in time to the beginning of the "visible" universe. The picture is admittedly fuzzy, but the coming years are bound to see ever increasing refinements.

It is ironic that with all the advanced technology required for obtaining the picture—the satellite, the delicate instrumentation it carried on board, the sensitive antennas on earth for receiving the signals transmitted from the satellite, and the powerful computers needed to deal with the masses of data—still the research team was confronted with the same problem that cartographers faced at the time of Columbus: how best to depict the surface of a sphere on a flat piece of paper. They could have chosen the two-hemisphere solution, or a Mercator view of the beginnings, or any of the many other compromises proposed over the years. They made a choice familiar to world-map watchers: a kind of slit-open, unrolled, flattened-out view of a spherical surface. Depending on what new features of the picture are filled in, we can expect other types of maps to follow.

The fact that we now have a picture of the curtain as it goes up helps to mitigate our disappointment at not being able to see behind or beyond it. In any case, we should not be too disappointed. First of all, we *are* able to see 99.99 percent of the way back to the very beginning. Second, on the basis of what we can observe directly, and what we know of the laws of physics, we can infer a great deal about what the universe must "look like" behind the curtain—at least until we get to the immediate vicinity of the big bang. Third, although we cannot receive electromagnetic radiation from the early stages of the universe, there are at least two possible methods for obtaining direct "messages" from that period. The first is by means of a type of particle called a *neutrino.* The current standard theory of the big bang calls for the release of enormous numbers of neutrinos—some hundred million for each atom in the universe. Unfortunately, the same property that allows the possibility of receiving neutrinos emitted at very early times in the history of the universe also makes them devilishly difficult to detect; neutrinos pass through ordinary matter even more easily than X-rays through clothes. A number of neutrino detectors have been built and they have proved successful in recording the arrival of neutrinos from the core of the sun and from distant supernovae. However, to detect neutrinos coming all the way from the origins of the universe would require a far more delicate apparatus than anything available today; that is a project for the next millennium. The second method for receiving direct information from the early universe may be much nearer to hand. In 1994 an international project called LIGO (Laser Interferometer Gravitational-Wave Observatory) was launched, with the goal of detecting "gravity waves." One can think of gravity waves as minute ripples in the curvature of space. The LIGO project is designed to detect these ripples coming from various sources, one of which would be the state of the universe in the first fraction of a second after the big bang. LIGO in the United States will operate in conjunction with VIRGO—named after the Virgo galaxy cluster—built by a French and Italian team near Pisa in Italy.

But these projects are still in the future. By the means already at our disposal, we have arrived at a picture that is perhaps in some ways only a crude first approximation with many details yet to be filled in, but that seems likely to hold up in its broad outlines. That picture—our current best picture of the retroverse—is of a hypersphere with a

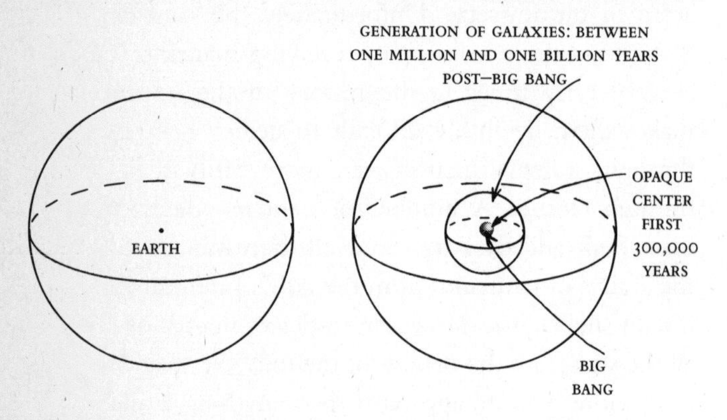

tiny ball excised. Our present instruments allow us to explore in some detail the outer boundary of that ball—the source of the microwave background radiation—and the results fit beautifully with the theoretical predictions. Surrounding the tiny ball centered at the big bang is a somewhat larger one—the birthplace of galaxies—and there our present instruments are woefully inadequate. We are now gradually obtaining a more detailed picture of the left half of our hypersphere, corresponding to galaxies within five or six billion light-years of the earth, and we have sent out fingers of exploration beyond into the right half. By all indications, millennium's end should bring us to the once unthinkable goal of a global picture—not the globe of the earth, but the globe of the universe.

Chapter VIII

Another Dimension

One factor that has remained constant through all the twists and turns of the history of physical science is the decisive importance of the mathematical imagination.

—Freeman J. Dyson

Albert Einstein was in every way larger than life. At a conference at Princeton University in 1951, organized to celebrate the one hundredth birthday of one of Bernhard Riemann's most original conceptions—the "Riemann surface"—participants were startled one morning to find Einstein, who was instantly recognizable, sitting in the first row of the lecture hall. He had come over from his office at the Institute for Advanced Study to say a few words of greeting. Most striking was the size of his head, appearing larger by half than those around him, even without the disorderly profusion of white hair surrounding it. Einstein spoke of his own great debt to Riemann and of how pleased he was that Riemann's seminal ideas continued to be explored and expanded. As it happened, the Riemann surfaces that were the subject of the conference are only distantly related to the Riemannian geometry upon

Einstein around the time of the Riemann surface centennial conference at Princeton in 1951 (American Institute of Physics Neils Bohr Library).

which Einstein founded his theory of relativity; three decades later, however, physicists found a significant use for Riemann surfaces in the "theory of strings"—a new attempt to understand the fundamental workings of the universe on the smallest subatomic level.

Part of what made Einstein so distinctive was a lifetime of defying convention and conventional wisdom, as seen in his trademark extravagant hair. He had strong convictions, and acted upon them, even when it took considerable courage to do so. Like Stephen Hawking, two generations later, Einstein combined a brilliant theoretical mind, most at home in the complexities and abstractions of modern physics, with a willingness to grapple with the concrete complexities of real life and to fight against heavy odds. In Hawking's case, the fight was against a crippling disease that hit him at the very start of his career. For Einstein, it was external politics that led him to become a pacifist opposing a militaristic Germany in World War I; years later he denounced Hitler in the face of grave personal danger. Also, both Einstein and Hawking fought back without bravado, but rather with a good-natured stubbornness, and both maintained throughout a remarkable sense of humor. Einstein's wry remarks and aphorisms were (and still are) picked up and quoted, both in and out of context, in support of every possible cause or belief. "I cannot believe that God plays dice with the universe" (in reference to the probabilistic interpretation of quantum theory). "Subtle is the Lord, but malicious he isn't." "Religion without science is blind. Science without religion is lame."

But as endearing as his personal qualities were, Einstein's place in history rests on his truly revolutionary insights into the nature of reality. For physicists, the twentieth century began in the year 1905, when Einstein published three celebrated papers, two of which overturned some of our most deeply held beliefs about the nature of the real world.

The first paper introduced what later became known as Einstein's "special theory of relativity." In it Einstein argued that we must abandon the notion of "simultaneity": for both practical and theoretical reasons, it is meaningless to assert that an event here and another event in the Andromeda galaxy happened "at the same time." Along with the loss of simultaneity, Einstein stated that such basic notions as measurements of length, velocity, and mass of an object were relative—that is, two observers could arrive at different, equally valid measurements, depending on their frame of reference.

Einstein's theory of relativity rapidly became a kind of metaphor, used to counter assertions of all kinds with the comment "Everything is relative." But Einstein's theory allows for no sloppiness or slippage. Each measurement of time, space, or mass can be carried out to whatever accuracy the instruments being used allow, and then precise mathematical laws, stated by Einstein, can be used to predict the results of measurements made by any other observer. Any sloppiness regarding Einstein's theory lies in the attempts to carry over the theory to the political, social, and moral domains.

Even more revolutionary in its own way than Einstein's special theory of relativity was another of Einstein's 1905 papers, in which he introduces the notion of a "photon" of light. That paper became one of the foundations of quantum theory, which, along with relativity, represented a totally new departure from earlier physical theories. In fact, although the invention of quantum theory is traditionally attributed to Max Planck, a case can be made that it was Einstein who first argued that energy was truly quantized, coming only in packets of a certain size. Planck appeared to view the "quanta" in his calculations more as a mathematical convenience than as physical reality.

If Einstein received too little credit for his role in originating quantum theory, posterity has more than compensated by crediting him with an idea that was not his: that of four-dimensional space-time as the fabric of the physical universe. It was Hermann Minkowski, one of the most original mathematicians of his time, who read Einstein's paper on special relativity and quickly realized that although measurements of space and time separately were relative to the observer, a certain combination of them was observer-independent. In a famous—if somewhat extravagant—pronouncement, Minkowski said, "Henceforth space by itself, and time by itself, are doomed to fade away into mere shadows, and only a kind of union of the two will preserve an independent reality."

To understand the issues underlying the four-dimensional approach to the universe, it is helpful to reexamine the *retroverse* from a more global point of view. By its very definition, the retroverse is constructed of everything observable at a given moment of (earth) time. But if we make those observations a year or a decade, or a century, from now, the picture of the retroverse will each time be different. Before Einstein, those pictures would have been said to describe "the same universe" at "later times." However, the inextricability of space and time made that description inadequate, even before the expansion of the universe was discovered. What we now realize is that each of those successive snapshots allows us to observe a part of the universe that simply wasn't visible before. The background microwave radiation received on earth a year from now comes from points one light-year further away than the sources that we now "see." We are literally expanding our horizons each year. In order to form a global picture we must envision a larger universe stretching out in space and time; each annual survey from earth reveals something like a thin slice of the universe. Our hope is to try to infer the shape of the entire universe from a number of these slices. It is somewhat like examining a few slices through an apple and trying to deduce the shape and composition of the whole apple. If the slices did not happen to go through the core, then we would have no way of knowing a core existed, much less what it looked like. Similarly, any attempt to "see" beyond the retroverse that we can actually observe in order to imagine the entirety of the universe must be at least partly speculative. The most common approach to forming a global picture is to assume that the universe is more like an onion than an apple, and each successive retroverse is like another layer peeled off the onion; successive layers or views of the universe may differ in detail, but resemble each other in overall structure.

The assumption that the universe is more like an onion than an apple is generally referred to as the *cosmological principle.* It expresses the notion that what we see from here on earth is typical of what observers anywhere else would see. It also expresses something more: that viewed on a large enough scale, the universe is smooth, rather than lumpy. In other words, the clumping of matter into stars and galaxies, with large spaces in between, is like the structure of a gas or liquid on a submicroscopic level, with most of the mass concentrated in the nucleus of each atom, and the atoms clustered in molecules. Moving from the atomic scale to the human scale, we perceive the gas or liquid as a smooth, undifferentiated substance. Similarly, the cosmological principle asserts that from a large enough point of view the individual galaxies will be like the atoms of an overall smooth substance—the substance of the universe.

Einstein had at least two reasons for believing in the cosmological principle. First, although it may be equally likely that there really are other reaches of the universe totally unlike our own, the chances of our guessing what those parts might look like are virtually nil. The probability that all parts of the universe resemble each other is vastly greater than that some other part is made purely of green cheese, for example.

Second, in the absence of any reason to expect a difference in unseen parts of the universe, it seems most probable that uniformity prevails. Our counterparts in the Andromeda galaxy would find that their retroverse overlaps to some extent with ours, but we have no reason to suspect that what they see and we do not, or what we see and they do not, is any different from what we both see.

The situation is much like our attempt to understand the shape of the earth at a time when only small parts of it had been explored. Based on the portions of the earth we knew, it seemed most likely that the remainder had the same overall spherical shape. In that case our belief was confirmed.

For these and other reasons, when Einstein expanded his special theory of relativity to the "general theory of relativity," in which he adopted Minkowski's description of a "four-dimensional space-time continuum" and tried to use his general theory to describe the whole universe, he decided to choose as a model of the universe a hypersphere evolving in time. He did so before the discovery of Hubble's law and its consequences, so that some of the reasons for his choice turned out not to be justified. However, five years later, in 1922, the Russian mathematician Alexander Friedmann constructed an improved version of Einstein's model. In Friedmann's model, as in Einstein's, the universe has a three-dimensional space component and a (one-dimensional) time component, where the space component at each instant of time has the form of Riemann's "spherical space" or "hypersphere." The most important difference between the two models is that Friedmann's allows for the expansion of the universe, later confirmed by Hubble. In other words, the size of the hypersphere increases as time progresses, and decreases to zero as time is reversed. The term "big bang" is often used to refer to the initial instant of time when the size of the universe is zero, but the big bang should be interpreted as a convenient abstraction rather than a physical reality. The physical reality, as best as we can determine today, is that the universe was once compressed into an inconceivably hot dense "primeval fireball." Beyond a certain point our physical understanding is incapable of encompassing all the possible effects and interactions of such a fireball. So we think of the big bang as an initial value of time, and speak of the physical universe only after the big bang. The picture then is relatively simple. If an "outside observer" with a video camera were to film the whole thing, then a succession of frames would reveal a succession of hyperspheres increasing in size.

Now, hyperspheres, like spheres, look the same from any point (which is why they are ideal if one believes in the cosmological principle). In drawing a picture of them, we can use an egocentric map with ourselves at the center, fully realizing that other observers would be free to draw their own maps centered at their own locations. So let us choose an electron attached to an atom at the center of the earth, and draw our map of each hypersphere with that electron as center, going back to about one second after the big bang; before that, electrons were being created and annihilated, so that we could not follow the history of our particular electron any further back.

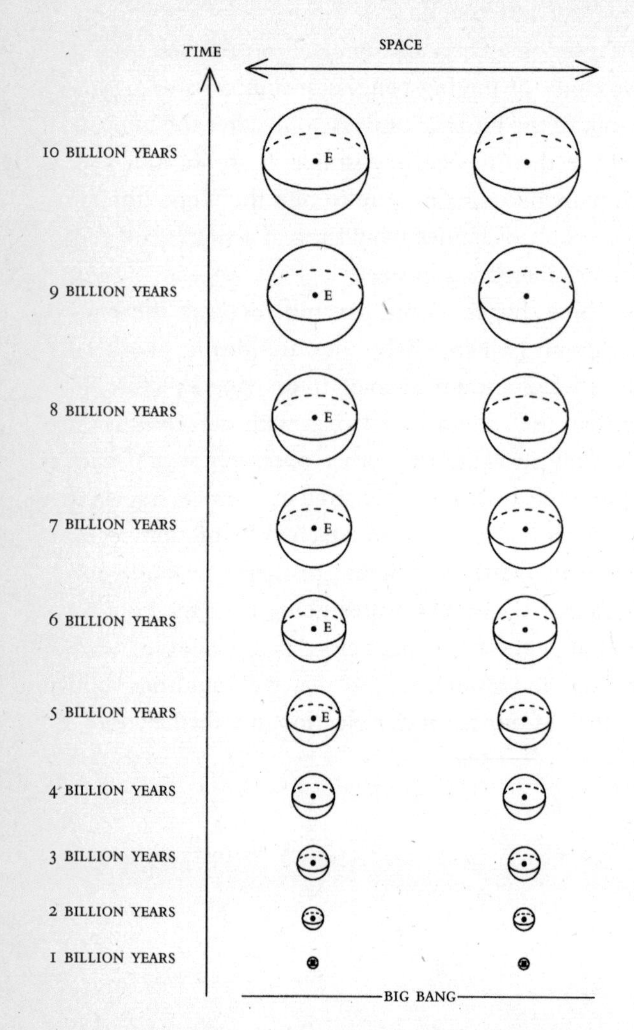

The first ten billion years in the life of a spherical universe. A series of snapshots at intervals of a billion years starting at the big bang. In each snapshot, space is in the form of a hypersphere, whose size grows steadily with time. ("E" is an electron at the center of the earth.)

At each instant of time after the big bang, all the particles in the universe are distributed on a hypersphere, which we depict in our usual manner as the insides of a pair of spheres. During the period from the big bang to the present time, the size of the hypersphere has been increasing steadily.

To fully appreciate this picture, remember that a sphere, or any other surface, may be represented on a map by considering circles centered at a given point. The curvature of the sphere or surface is reflected in the way the length of those circles grow with distance from the center. The curvature of space, according to Riemann, may be defined in a similar manner. We have now taken the next step, and described a universe in which all particles at a given "distance" from the big bang lie on a (three-dimensional) hypersphere, where "distance from the big bang" is simply measured by time. The resulting global picture can only be described as something four-dimensional; we cannot depict it in its entirety, but we can nevertheless describe it fully by giving the size and shape of the hyperspheres at each time *t* after the big bang.

The *idea* that space and time combined may be considered as something four-dimensional is a very old one; it is explicitly mentioned in an article on "dimension" in the famous French encyclopedia published in 1764. There, however, it is simply a throwaway idea, not followed up in any fashion. Not until Minkowski's interpretation of special relativity was the mathematics of four-dimensional space-time spelled out in a way that could be used to solve concrete physical problems. In his formulation of the general theory of relativity, Einstein ties the knot of space and time still more tightly by introducing the notion of curvature in space-time, showing explicitly how to deduce the effects of gravity from the curvature.

It was here that Einstein drew his inspiration from Riemann. Riemann's vision was not limited to three dimensions; it extended to four and higher dimensions, where Riemann introduced the notion of curvature and gave explicit equations for calculating it. Einstein's genius lay in seeing first that Riemann's equations could also be used in space-time, and second, that the geometry of space-time then influences the physics. The latter notion was truly revolutionary, since in all previous scientific theories, space was a passive background—the stage upon which the action took place. In Einstein's formulation, both radiation and material objects move along paths determined by the geometry of space-time. "Gravity is geometry," one might say. Although Einstein introduced that idea early in the twentieth century, it retains today much of its mystery. It might help to remember that Newton's explanation of gravity was considered equally baffling at the time it was proposed, and it took at least as long to be accepted. In fact, the closer one looks at Newton's law of gravity the more bizarre it appears: any two objects exert a force of attraction on each other, and that force is somehow transmitted instantaneously across the vastness of empty space, from sun and moon to earth, from star to star, from galaxy to galaxy. Many reputable scientists at the time dismissed the idea as "voodoo physics" and refused to take it seriously. Newton himself explicitly disclaimed any understanding of the physical mechanism involved, and explained that all he did was to provide the mathematical laws from which one could compute the motion of any object under the influence of gravity; he would leave it to future generations to unravel the secret of how the mysterious "force" of gravity really operates.

Looked at from that perspective, Einstein's notion of warped space, favoring certain trajectories, may not seem so outlandish.

To see why gravity might indeed be curvature in disguise, we need only recall one of the key properties of curvature that we have described for surfaces and for space. In two dimensions, we measure the length of circles centered at a point, and see whether the length increases more slowly, more quickly, or at the same rate as in the plane (telling us whether the curvature is positive, negative, or zero, respectively). In three dimensions a similar computation, using the size of spheres centered at a point, gives a measure of curvature. In four-dimensional space-time, we have hyperspheres at a given "distance" from the big bang, and they grow in size as the distance increases. The *rate* at which they grow gives, then, a natural measure of the curvature of space-time. On the other hand, the rate at which the hyperspheres grow depends on the braking action of gravity in slowing down the expansion of the universe. Thus, increased gravity, slower growth of the hyperspheres as time advances, and greater curvature are all ways of describing the same phenomenon.

Once we have the picture, we can break free of our physical confines of looking backward in time and let the

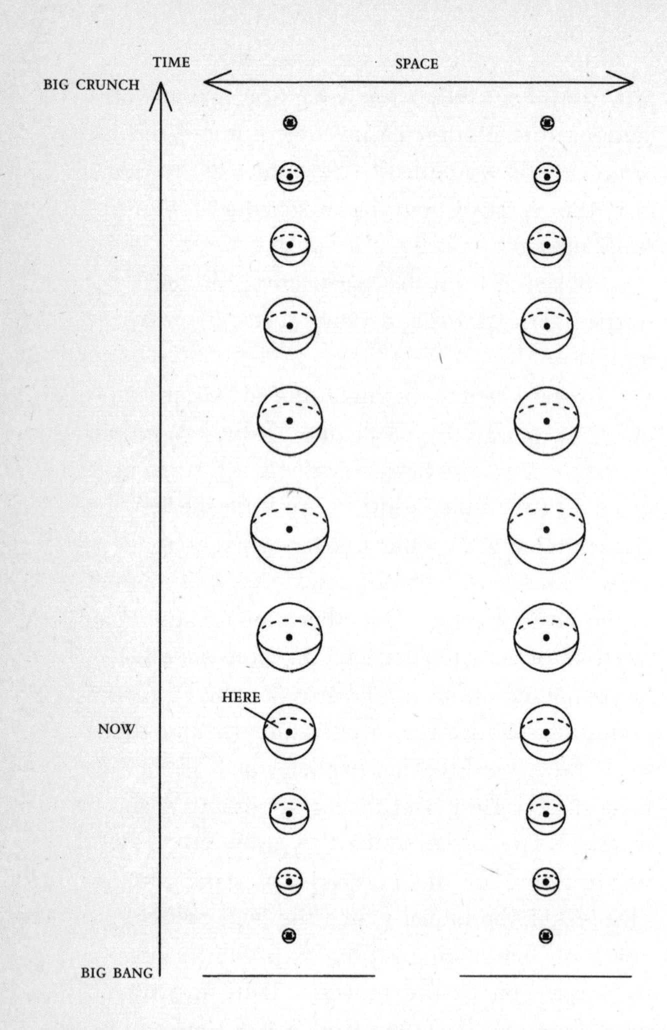

Map of the universe as a four-sphere. At each instant of time after the big bang, space is a three-sphere, depicted here as the interior of a pair of spheres with the two boundary spheres representing the same sphere in space. The size of the three-sphere grows until it reaches a maximum size (sometime in the future) and then shrinks back down toward the big crunch.

mathematics predict the future of the universe, just as Newton's laws allowed a detailed description of the future course of the solar system. As time evolves, with the universe expanding, distances between galaxies grow and gravitational forces weaken. That means that the space-time curvature diminishes and successive hyperspheres grow at a slower rate. Two possibilities emerge. The first is that the hyperspheres continue to grow indefinitely, although at an ever decreasing rate. The second possibility is that they reach a maximum size and start to contract, in exactly the same fashion as the parallels of latitude on the earth, starting at the North Pole, grow till they reach their maximum size at the equator, and then begin contracting again toward the South Pole. If the universe does someday enter such a contracting phase, then distances between galaxies would start to decrease, gravitational attraction would increase, curvature would increase, and successive hyperspheres would shrink ever faster, eventually contracting back down to a single point, known popularly as the "big crunch." We could then draw a map of the entire universe as a succession of hyperspheres growing in size for the first half of the life of the universe and contracting for the second half. All space-time would then form a kind of super-hypersphere: a four-dimensional object that mathematicians call a "four-dimensional sphere," or "four-sphere" for short.

Our current knowledge of the universe is not sufficient to decide whether there is to be a big crunch in our future or whether the universe will go on expanding forever. But in either picture, it is worth taking a last look back and seeing how our (three-dimensional) retroverse fits into the big picture of the entire (four-dimensional) universe. Since the early part of the universe—the initial expansive phase—looks much the same in the two cases, we can draw a map of the universe as a four-sphere and sketch in the current retroverse.

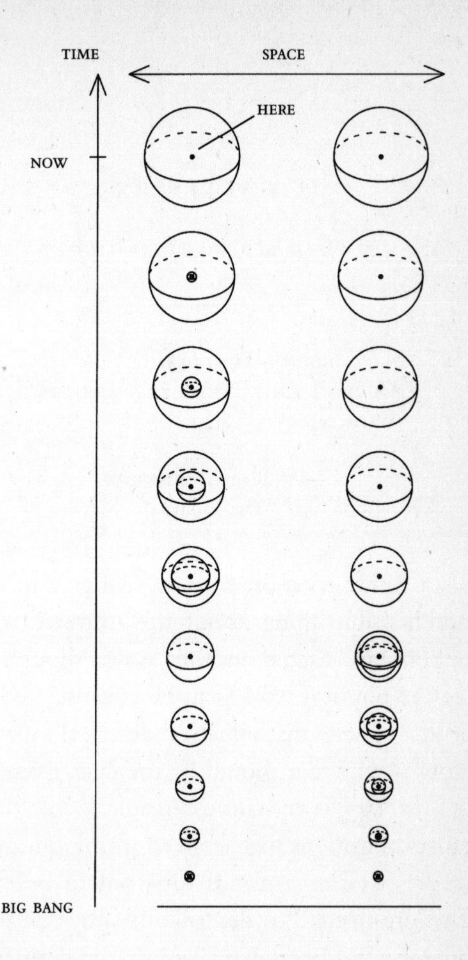

The current retroverse as part of the universe from the big bang to the present. The universe is shown as a three-sphere growing in size with elapsed time from the big bang. The observable universe, or retroverse, is depicted as a succession of (ordinary) spheres growing in size (from top to bottom) as we look back from the present, starting out in the left-hand part of the three-sphere, and gradually increasing in size, until it moves over to the right-hand part and shrinks down to the central point at the moment of the big bang.

CHAPTER IX

A Galaxy of Shapes

MIGHTY IS THE CHARM
OF THOSE ABSTRACTIONS TO A MIND BESET
WITH IMAGES.

—William Wordsworth,
"The Prelude"

One of Einstein's most famous sayings was: "The most incomprehensible thing about the universe is that it is comprehensible." Stated another way, although many aspects of the physical world can be encapsulated in simple laws or in a concise mathematical description, what we do not know is why that should be the case. Even more difficult to explain is the almost magical way that certain mathematical notions that seem to spring out of sheer invention from creative minds turn out to be exactly the tools that are needed to describe the physical world. This phenomenon has been described by the eminent twentieth-century physicist Eugene Wigner as "the unreasonable effectiveness of mathematics in the natural sciences." A striking example is the theory of conic sections—the ellipse, parabola, and hyperbola—first studied for no apparent practical reason by Greek mathematicians around 400 B.C. That theory did not find an application in science until two thousand years later when Kepler realized that the shape of a planet's orbit around the sun is an ellipse. Kepler's discovery was further elaborated by Newton to include comets and other objects entering the solar system, whose orbits could be an ellipse, parabola, or hyperbola. Newton also showed that the shape of the earth itself was ellipsoidal rather than spherical.

Another shape from antiquity had an even longer wait before finding a practical application in science, in this case, in the field of chemistry. In 1985, Harold Kroto and Richard Smalley collaborated to determine the molecular structure of a newly discovered form of carbon, in which sixty carbon atoms are somehow linked together to form a single molecule. The molecule's structure was initially a great mystery. It turned out to have a shape that was described in the third century B.C. by Archimedes, consisting of a symmetrical arrangement of pentagons and hexagons, put together in a design now familiar around the world as

A soccer ball's symmetrical arrangement of pentagons and hexagons, which solved the molecular structure of the sixty-atom molecules known as "buckyballs."

the pattern on a soccer ball. The use of similar designs by Buckminster Fuller in the construction of geodesic domes led Kroto and Smalley to christen the new molecule and similar ones analyzed later *buckminsterfullerenes,* a term that has since been mercifully and humorously shortened to "buckyballs." They are now the subject of intensive research, as practical applications loom on the horizon.

The mathematical constructions discussed so far, from curved space to Riemann's hypersphere, have already proved their value in explaining and understanding our universe. More recent creations of the mathematical imagination, dating mostly from the twentieth century, have not yet had their applicability to modern science fully established, although there are already strong indications that they are forthcoming.

If there is a single key to the way new mathematical constructs are created, it is through the process of abstraction. One of the most familiar and most important examples of that is the notion of "number." Numbers themselves do not occur in nature. It was a momentous realization that although you may not be able to add apples and oranges, you *can* add the *number* of apples to the number of oranges and obtain the correct total number of pieces of fruit. Furthermore, the rules for addition are universal, quite independent of the particular objects to which the numbers originally referred. The difficulty of making the step from numbers of concrete objects to an abstract number is clearly evident in the fact that even today, certain languages, such as Japanese, use different words to describe the same number when attached to different kinds of objects.

Abstraction works in a variety of ways. First, it carries the power of universality, allowing a single rule to apply in very different circumstances. The fact that three times five equals fifteen applies equally when calculating the total cost of three tickets at five dollars each, or when trying to determine how big an area one needs to varnish in order to cover the surface of a three-by-five tabletop. The banality of the arithmetic tends to obscure the act of genius that abstracted the notion of a number from the jumble of particular numbers of particular objects.

A second advantage that abstraction offers is that it often brings clarity to what may be a confused situation. The notions of "point" and "line" in Euclid's writings, for example, are rendered much more transparent and subject to far simpler rules than the dots and dashes in real life from which they are abstracted. The danger, of course, is that conclusions we reach by applying the simpler rules to our abstractions may not be valid when we try to apply them back to the original real-life objects. However, the remarkable fact alluded to by Wigner in his "unreasonable effectiveness of mathematics" statement is how often those conclusions are right on the mark.

The third big advantage of abstraction is that it provides us with great freedom to let our imaginations roam, permitting us to devise new and alternative versions of reality—versions that may or may not correspond to something in the real world. Numbers, for example, had been in use for thousands of years when the notion of a "negative number" was tentatively introduced. The idea initially met with great resistance, since it represented a different level of abstraction from ordinary numbers. The number five could be clearly understood in relation to collections of five objects or lengths of five units. Calling "negative five" a "number," when it did not correspond to anything concrete, took the construct one step further. Today, we are so used to accepting and working with negative numbers that it is hard for us to realize how problematic they were when first introduced.

It should be no surprise that those parts of mathematics based on abstractions taken directly from the real world, such as numbers and euclidean geometry, are useful and applicable to real-world problems. What Wigner referred to in his quote about the "unreasonable effectiveness of mathematics" is the applicability of the most abstruse realms of mathematics, in which abstractions have been piled on top of abstractions. For example, after the notion of a negative number was gradually absorbed, a still more improbable-sounding notion was introduced—that of a "number" whose square was a negative number, in contradiction to the basic rules of arithmetic. These new objects were called "imaginary numbers," and they met with even greater resistance. And yet, when properly interpreted and understood, they took their place as a standard tool in a mathematician's kit and became an indispensable component for large parts of physics and engineering.

Let us now describe some abstractions of the more adventurous kind. The first of those is what is sometimes called an "abstract surface" and sometimes, in more strait-laced terms, a "two-dimensional manifold." Perhaps a better term would be a "designer surface," since, like "designer drugs" that may not occur naturally but are put together to our own specifications, a designer surface is one that we define into existence as an abstract object. It may or may not have a real-world counterpart.

There are several different ways that one can design such a surface. One of the simplest and most useful is to take a familiar shape, such as a rectangle, and decree that two of the sides of the rectangle are really the same side on the new, imagined surface. It is as if the two sides are simply glued together. For example, if we define the two vertical sides of a rectangle to coincide, we obtain an abstract surface which is the precise playing field for a number of the early video games, such as Pac-Man, in which a figure moving off the right side of the screen immediately reappears on the screen on the left. Long before video games, the same idea had been applied to a certain kind of chess, where all the ordinary rules of chess applied, except that the left and right edges of the chessboard were treated as the same line, so that a chess piece could be moved across the right-hand edge and immediately reappear on the left. That game was termed "cylindrical chess," for the simple reason that the abstract surface on which it is played corresponds to a familiar real-life surface, the cylinder. (If we cut a rectangle out of a piece of paper, and bend it around so that the vertical edges

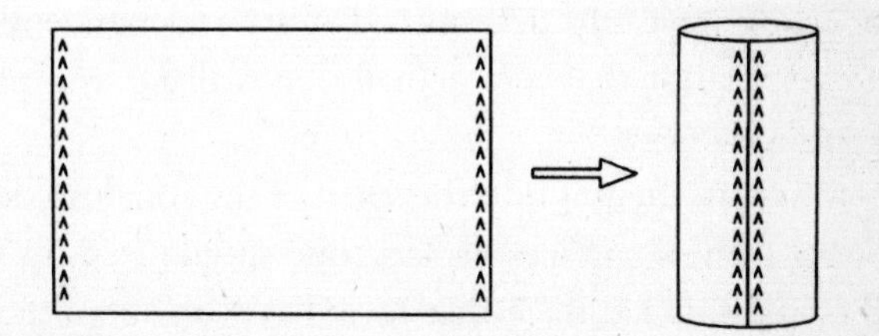

come together, it forms a cylinder, which is an exact model of the surface that we designed abstractly.)

A slight variation on the same idea produces a famous surface that was not created until 1858: the *Moebius strip.* Again one starts with an ordinary rectangle and decrees the two vertical sides to be the same line on the surface, but this time with the directions reversed: the top of the left-hand side matches the bottom of the right and vice versa. Does this imaginary surface of Moebius correspond to a surface in the real world? The answer in this case, perhaps surprisingly, is "sometimes." If we start with a long, thin rectangle on a sheet of paper, we can cut it out, and bend it around so that the two narrow edges match, top to bottom as prescribed, by making a half-twist before bringing the edges together. On the other hand, if the rectangle is a square, or close to being square, then we are unable to twist it around to bring two sides together, top to bottom; in that case we have a perfectly good abstract Moebius strip that has no real-life counterpart. Since there would be no difficulty whatever in programming a video game to follow the specifications for a Moebius strip, regardless of the shape of the initial rectangle, Moebius strips of all sizes and shapes enjoy the same virtual reality. From a mathematical point of view, all Moebius strips are created equal, and their geometric properties are easily determined quite independently of whether or not they can actually be realized as a physical surface.

Once having got into the spirit of this, one can design all sorts of new surfaces. In fact, one simple step up from the cylinder is a surface that *never* has a real counterpart.

But before taking that step, it may be useful to pause a moment to touch on the question: Are "designer surfaces" truly "designed" or are they "discovered"?

That question is part of the larger, ongoing debate as to whether mathematicians "invent" or "discover" the objects that they study: numbers, fractions, irrational numbers, imaginary numbers, circles, spheres, hyperspheres, pseudospheres, and so on. The Moebius strip is a good case in point. It is generally viewed as the quintessential artificial object, made by twisting and gluing a strip of paper, from which it derives some of its paradoxical properties. For example, if you wish to paint a yellow line down the center of one side and a blue line on the other, you discover that it cannot be done: there is no "other" side. Starting anywhere with your yellow line, and continuing down the center line until you come back to where you started, you find that you have traversed both sides of the surface without ever crossing an edge. In other words, the Moebius strip is in some sense a surface with only one side. The same is true of the two "edges" of the strip, which upon closer examination turn out to be a single continuous edge. That observation is dramatically confirmed when you try to cut the strip in two along the center line, and find that you are left with a single connected strip of half the width of the original.

This strange behavior is generally attributed to the fact that the Moebius strip is something concocted with scissors and glue, in contrast with "real" surfaces, such as a sphere made by a soap bubble or the ellipsoidal surface of the earth. But in fact, if you would like to witness nature making a Moebius strip, simply bend a wire in the form of a double loop and dip it into a bowl of soapy water. When you withdraw the wire from the soap solution, you

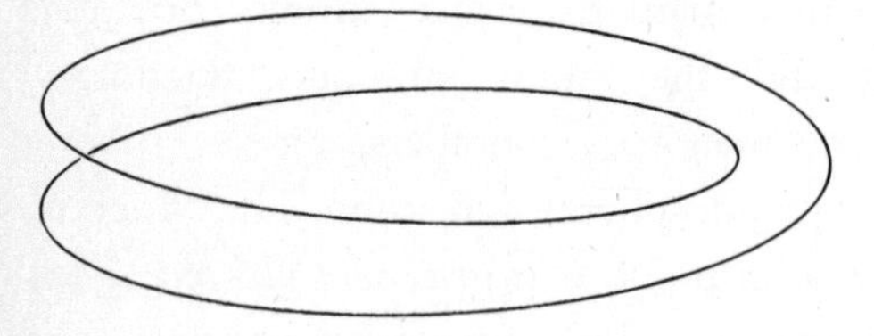

A double loop.

typically find a soap film spanning the wire in two pieces—one going around the edge and the other across the middle. If you pop the film across the middle (using a dry finger, for example), what is left is precisely a Moebius strip, with all of its paradoxical properties. Using a liquid plastic instead of soap, one can create a permanent nature-made Moebius strip, as real as any surface you might care to name. And so August Moebius in one sense "invented" the surface that is named after him, and in another sense "discovered" a surface that already existed—at least potentially—in the physical world.

Another design that mathematicians came up with begins again with a rectangle. In this case, we decide that the right and left edges are one and the same line and *also* that the upper and lower edges are the same line. Once again, we can construct a video game proof of the virtual reality of such surfaces by programming a space adventure in which rocket ships moving across the right edge of a video screen reappear on the left and those crossing the top edge reappear on the bottom. To test the actual reality of such surfaces, one would start by bending a rectangle around to match the two vertical edges, forming a cylinder (Fig. 1). Then the original top and bottom edges of

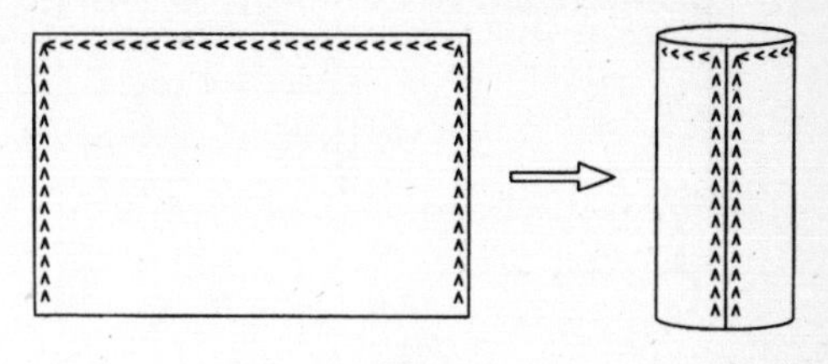

Fig. 1

the rectangle correspond to a pair of circles on the top and bottom of the cylinder. In order to match the top and bottom, we would have to somehow bend the cylinder around, without stretching it, to bring the two circles together. We sense intuitively, and can prove mathematically, that such a construction is not possible. Nevertheless, as an abstract surface, this particular one has had a long and vigorous life. It goes by the name of a *flat torus.* The word "torus" reflects the fact that the surface is in important ways the same as the surface of a doughnut. For example, a horizontal line connecting the two vertical edges of the original rectangle corresponds to a circle on the surface, as is clear if we take the first step in trying to realize the surface by bending the rectangle around to form a cylinder (Fig. 2). In the same way, a vertical line connecting the top

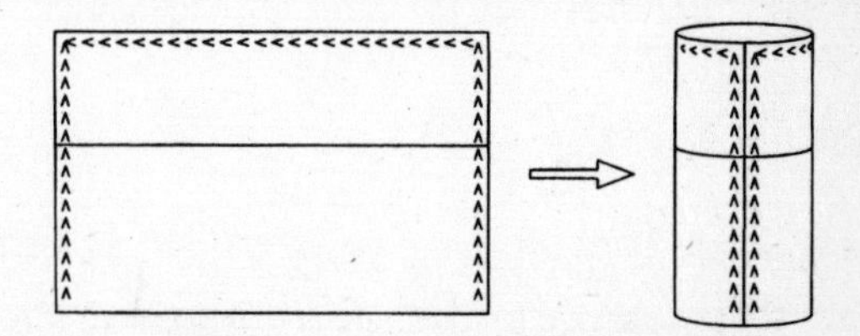

Fig. 2

of the bottom of the rectangle forms a circle on the surface (Fig. 3). (We could start by bending the rectangle

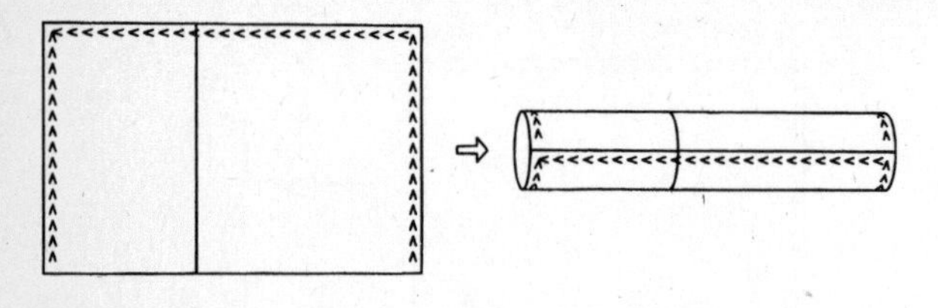

Fig. 3

around the other way, joining the top to the bottom.) Since a horizontal and vertical line meet at a single point, we have two circles on our surface that cross at a single point (Fig. 4). That can easily happen on the surface of a

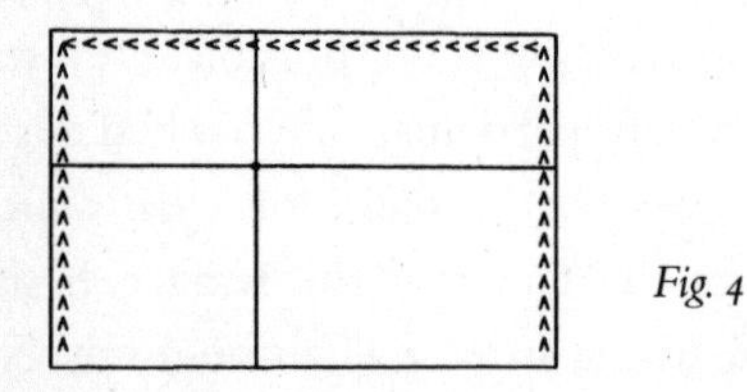

Fig. 4

doughnut, but not, for example, on the surface of a sphere. (A circle on a sphere divides the sphere into two parts, and

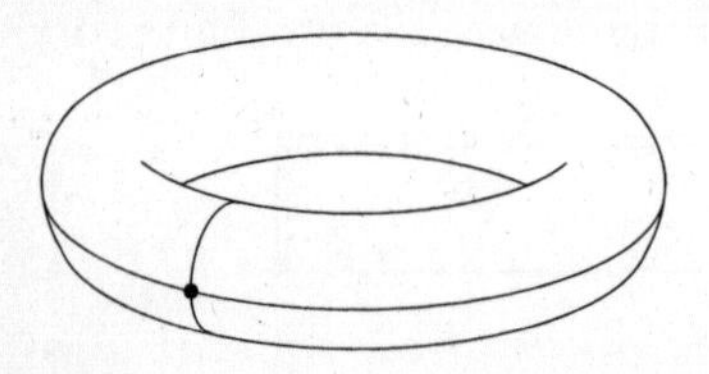

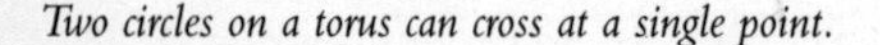

Two circles on a torus can cross at a single point.

a second circle that crosses the first circle once must cross it a second time.)

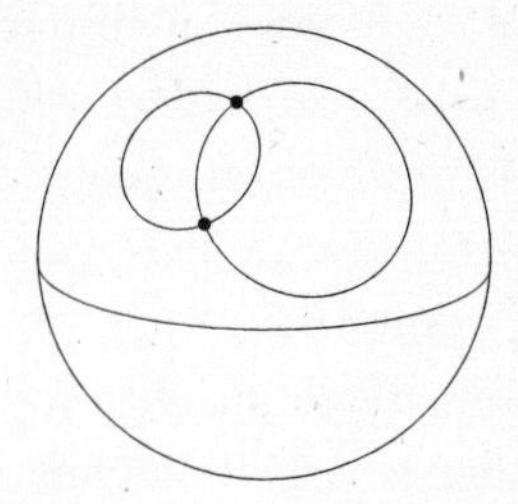

Two circles on a sphere that cross at one point must cross back again at a second point.

In contrast to an ordinary torus, or doughnut, which is curved, a flat torus, as the name implies, is not. The rules of ordinary euclidean geometry apply to triangles and other figures on the surface. It is a curious fact that even though a flat torus does not exist as an ordinary surface in space, it *is* found among the denizens of Riemann's hypersphere. That fact was first noticed by the English mathematician William Kingdon Clifford. As a result, the flat torus inside a hypersphere is called a Clifford torus.

These few examples give some inkling of the enormous variety of surfaces that mathematicians have designed, once they realized that such surfaces could be just as interesting as the surfaces found in nature. Other examples of particular interest go by the names of the "cross-cap" or "projective plane," the "Klein bottle," and the "Klein curve of genus three." The latter two are named after one of Germany's leading mathematicians, Felix Klein, who revitalized geometry toward the end of the nineteenth century.

One advantage of the rather cumbersome name "two-dimensional manifolds" for abstract surfaces is that it points the way to the next logical step of abstraction or generalization: "three-dimensional manifolds," called "three-manifolds" for short. Once again, the best description is by way of examples.

Picture an ordinary room with four walls, floor, and ceiling. Picture then a 3-D holographic video game in which images of three-dimensional spaceships move about the room and occasionally run into one of the walls, upon which they reappear at the corresponding point on the wall directly opposite. In the same way, they can pass through the ceiling and reappear at the point directly below on the floor. All inhabitants of this specially designed world follow the same pattern. The virtual reality that they then inhabit is what we call a "three-torus." It is a three-dimensional manifold that is the exact analog of the flat torus in two dimensions.

Another example of a three-manifold is one already familiar to us, the hypersphere of Riemann, which also goes by the name of the "three-sphere."

Before following this path any further into the jungle of abstractions, one might ask if these self-designed three-manifolds have counterparts in reality. The answer is that indeed they may. Twentieth-century cosmologists have been examining a whole range of three-manifolds in the hope of finding one that provides the closest fit to the shape of the universe shortly after the big bang. Both the three-sphere and the three-torus are possible candidates.

In fact, the three-torus provides a possible way out of the old dilemma of choosing between a finite and an infinite universe. The modern version of the dilemma is worth describing in some detail.

Before tackling the structure of the universe we must first go one step further into the realm of abstraction, from three-manifolds to four-dimensional manifolds, or four-manifolds. The basic tenet of Einstein's general theory of relativity is that whenever gravity comes into play—whether we are computing the orbit of an artificial satellite around the earth or a planet around the sun, or a pair of binary stars about each other, or the neighborhood of a black hole, or the structure of the entire universe—we must set up our system in the form of a four-dimensional curved space-time, where gravity is encoded in the curvature.

If we accept the premise of the big bang—which various forms of physical evidence seem to confirm—then we are able to disentangle the space and time components of space-time by measuring time as beginning with the big bang and picturing space as (necessarily) a three-manifold at each instant of time after the big bang, evolving as time advances. One such scenario is the original model of the universe proposed by Einstein where space is in the form of a three-sphere. In that model, the curvature of the universe is positive. On the other hand, it is also possible that the curvature of the universe is zero. If so, we are led to a model of the universe in which space at any instant after the big bang, instead of being a three-sphere, is the much more familiar euclidean space. As time advances, space expands, with galaxies spreading further and further apart; that process goes on forever in what is known as an "open universe." In contrast, a universe in which both space and time are finite is called a "closed universe." The basic model of a closed universe is the one proposed in 1922 by Alexander Friedmann. Friedmann's model, like Einstein's, is based on Riemann's spherical space, but in Friedmann's universe, space goes through a period of expansion, with the rate of expansion gradually decreasing until at some point in the distant future it stops altogether, at which time gravitational attraction begins to force space to contract until it at last collapses in an implosion referred to as the big crunch.

The "open universe" of modern cosmology is not all that different from the ancient view of the universe as infinite in extent, starting at some time in the past and going on forever. The only real difference, in fact, is that the universe is no longer static, but rather constantly expanding as galaxies move further and further apart. The difficulty with the modern picture of the open universe is that it suffers on a cosmic scale from the image—like that of Athena springing full-grown from the head of Zeus in the ancient legend—of appearing at once, full-blown, and infinite in extent at the instant of the big bang. As an abstract mathematical model of a possible universe, it is perfectly feasible, and there is no physical evidence one way or the other to rule it out. Nevertheless, many of those throughout the ages who thought deeply about the question came to the conclusion that an infinite universe made greater sense as an abstract model than as a physical reality.

We could also try to fashion a model in which we

have the best of both universes: from the open universe incorporating the feature of infinite time, with space continually expanding (though at a slower and slower rate), while at any given instant of time, space is finite in extent, as in a closed universe. The problem is that if we accept certain natural assumptions about the nature of space-time, and if we apply Einstein's equations of general relativity, then in the case of positively curved space-time we are led to the spherical model, finite in space and time, whereas for zero-curvature space-time, we have time going on forever and space at any given instant of time being flat, or euclidean.

But there is a way out of this dilemma. One of the gifts of the mathematical imagination is the construction of a three-manifold that is flat, with zero curvature, like euclidean space, but finite in extent: the three-torus. Just as any portion of the two-dimensional flat torus is indistinguishable from a part of the euclidean plane, so any part of a three-torus cannot be told apart from a corresponding part of ordinary euclidean three-space. But the three-torus is finite in extent, and if one keeps going long enough in some direction, one ends up eventually back in the vicinity of where one started, just as happens on a three-sphere.

And so we have a third alternative to the closed universe and the open universe: the half-open universe, finite in size, but extending indefinitely in time. On the basis of current knowledge, we have no way of preferring one model over the others.

There are still more abstract manifolds and other possible models of the universe. An important set of models evolved out of the "hyperbolic plane": the realm of non-euclidean geometry according to Lobachevsky and Bolyai. A famous theorem of Hilbert tells us that there is no actual surface in ordinary euclidean space corresponding to the hyperbolic plane. But as an abstract surface, the hyperbolic plane has played a seminal role in many areas of mathematics and physics.

In the hyperbolic plane one has geodesic triangles, quadrilaterals, and other polygons, just as in the ordinary euclidean plane. With those polygons we can play the same kind of game we did with a rectangle to produce abstract surfaces like the Moebius strip and the flat torus, to obtain a whole array of fascinating new abstract surfaces. Some of them correspond to real surfaces in space, such as the pseudosphere, while others do not.

Lobachevsky did not limit himself to two-dimensional non-euclidean geometry, but considered three-dimensional space as well. The result is "hyperbolic space"—a three-dimensional manifold with fixed negative curvature in the sense of Riemann. In hyperbolic space there are rooms with many walls, and just as we fashioned a three-torus out of a room in euclidean space, we may design many new three-manifolds out of those rooms in hyperbolic space. The study of those manifolds, called "hyperbolic manifolds," has been one of the most actively pursued areas of geometry during the second half of the twentieth century.

Cosmologists have suggested, in addition to models of the universe with positive and zero curvature, other ones with negative curvature. These models turn out also to be open universes, extending forward indefinitely in time, and the shape of space at any given time after the big bang is a hyperbolic manifold: a negatively curved three-dimensional manifold. The usual assumption is that space is infinite in extent, consisting of all of hyperbolic space, but another equally valid option is one of the many finite-sized hyperbolic manifolds that mathematicians have uncovered.

And so Riemann's original vision has evolved over the past century and a half. Starting with an attempt to find a mathematical language to describe the shape of space, he created the notions of curved space and a three-dimensional manifold. He then generalized those notions to manifolds of four and more dimensions and explained what would be meant by "curvature" in that context. His ideas were seized upon by mathematicians, who constructed many examples of such manifolds and studied their properties, creating a whole field of mathematics known as Riemannian geometry. During the twentieth century the development of Riemannian geometry has gone hand in hand with attempts to unravel the secrets of the universe, as abstract manifolds conjured up by mathematicians continue to provide possible models to test against actual observations.

Even as physicists were gradually getting accustomed to thinking and working in four and more dimensions, mathematicians were moving into domains that seemed far more strange—even bizarre. In 1918, the year after Einstein presented his model of the universe as a positively curved four-dimensional space-time, the German mathematician Felix Hausdorff suggested a possibility never before considered: *fractional* dimension. The importance of

Hausdorff's new notion was not immediately apparent. As the usefulness of his ideas grew, however, they gained gradual acceptance and finally an enthusiastic embrace as an integral part of mathematics. But it was not until 1975 that Hausdorff's notion of fractional dimension became known outside the world of mathematics. Benoit Mandelbrot, a mathematician working at IBM, wrote a book on objects with fractional dimension in which he coined the term "fractal" to refer to them, and made the key observation that fractals were not just a figment of mathematicians' overheated imaginations, but in fact were more the rule than the exception in nature. During the past twenty years, applications of fractals have ranged from chemistry and metallurgy to the design of imaginary landscapes in films.

The original problem that Hausdorff wrestled with was one that had plagued mathematicians down through the ages: how to determine the area of a surface. The intuitive idea of area is clear enough. It is simply a question of how much paint, for example, you need to cover a surface. If the surface consists of the walls and ceiling of a standard rectangular room, then the area of each rectangular portion is simply the width times the height. For an octagonal room with an octagonal ceiling the calculation is a bit harder, but still pretty straightforward. But what if you need enough paint to cover a circular dome? Or the elliptical domes of various Italian cathedrals? Or one of Gaudí's free-form creations in Barcelona? Or the exterior of an ocean liner? The further and further one strays from simple shapes, the thornier the problem of area becomes. During the nineteenth century a number of ways had been suggested to determine the area of general surfaces, but with each new suggestion, examples were constructed to show the weakness of the proposed method.

A simpler but similar problem was that of determining the length of a curve. The answer had seemed pretty well in hand, until a very strange phenomenon appeared toward the end of the nineteenth and the beginning of the twentieth century. It turned out that the closer one looked at curves and surfaces, the more complicated they became. The final blow to any naive belief that one could rely safely on geometric intuition was the discovery of examples where it was not even clear whether the object in question was a curve or a surface! One such example, discovered in 1890 by the Italian mathematician Giuseppe Peano and known as the "Peano curve," appears from its construction to be a curve, but in fact passes through every point inside a square. Should one talk about its length or its area?

The example that became the archetype for the future theory of fractals was discussed by the mathematician Helge von Koch in 1904 in the first volume of a new journal devoted to mathematics and astronomy. It is generally called the "snowflake curve" or "Koch curve." It reminds one of a snowflake, both because of its six-sided symmetry and its lacy boundary.

The method for constructing Koch's abstract snowflake is closely analogous to the process that forms a real-life snowflake. It is a process of accretion, starting with a central kernel, which "grows" a symmetric set of six crystals, each of which then grows further crystals, and so on. Koch's abstract snowflake may be pictured in exactly the same fashion, where the add-on "crystals" are all in the simplified form of equilateral triangles. At each stage, the add-on triangles have sides one-third the length of those in the previous stage. The central kernel may be thought of as a single large equilateral triangle, which does not yet have a six-sided symmetry. But after the first stage, when an equilateral triangle of one-third the original size is attached to the center of each side of the central triangle, one obtains a regular six-pointed star. At the next stage, each straight side of that star "grows" another equilateral triangle, one-third the size of the previous ones, and the process repeats itself indefinitely. The end result is

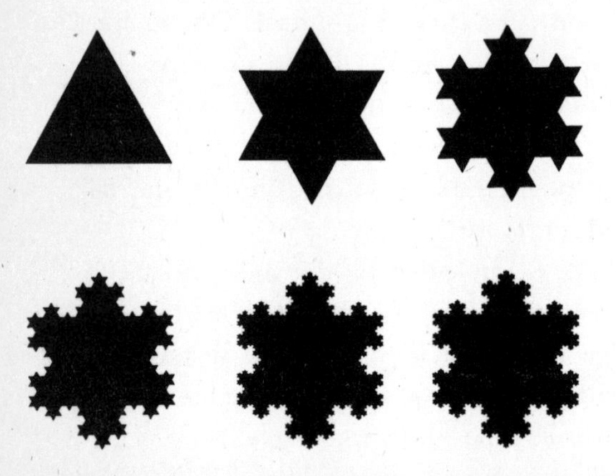

The first six stages in the "growth" of von Koch's snowflake.

the Koch snowflake, and it is the filigreed boundary of that snowflake which is the "snowflake curve."

The question that sparked the line of investigation leading to fractional dimensions was how to determine the total length of a snowflake curve. To answer that question, we might use one of the pieces of the curve itself as a "ruler" to measure the total length. Since the full curve is made out of twelve identical pieces, if we use one of those pieces as a measuring stick, then the total length would be twelve "units." But if we try to determine the length of just the upper third of the snowflake

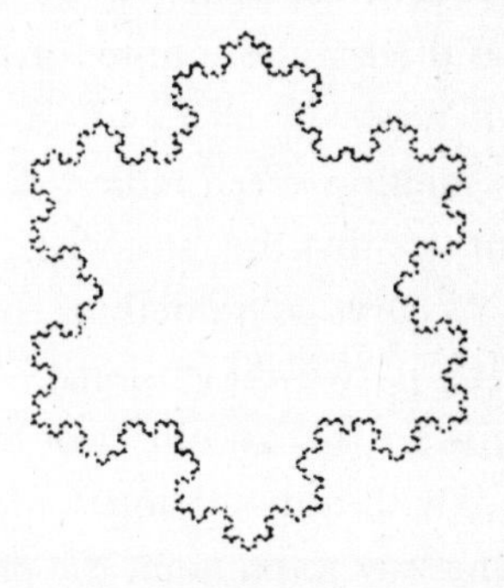

The snowflake curve.

curve, then a problem arises. Since it consists of four identical copies of our "measuring stick," the upper third should have a length of four units. But it can also be calculated in a different way: if we enlarge our measuring stick by simply scaling it up by a factor of three, then we get an identical copy of the *whole upper third* of the snowflake curve. That would indicate that the length of the upper part of the curve is three times the length of

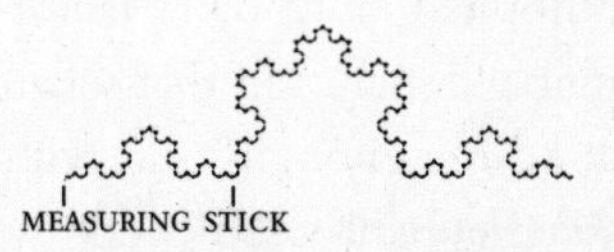

The upper third of the snowflake curve.

the measuring stick, or that its length is three units, rather than four.

There are two possible solutions to this apparent paradox. The first, which would have been the standard answer in the nineteenth century, says that the length of the "measuring stick" is itself infinite, so that the length of the upper third of the snowflake curve is also infinite; whether we consider it as three times infinity or four times infinity makes no difference.

Hausdorff's solution was a radical departure. He said that the problem was that the "snowflake curve" is not in fact a "curve." A curve is something one-dimensional, whereas a surface is two-dimensional. The "curve" designed by Koch is actually an object of fractional dimension, he claimed, with a dimension somewhere between one and two. One way to pinpoint the precise dimension is to think in terms of scaling.

When the Statue of Liberty was built, the first step was the construction of a small-scale model. Building the statue to scale from the model in one step would have been a very risky operation. The sculptor decided to construct a number of intermediate models, scaling the statue up by a modest amount at each step. One reason for caution is that each time the scale is doubled, the volume—and therefore the weight—goes up by a factor of eight. On the other hand, the strength of the supports, which depend on cross-sectional area, is multiplied only by a factor of four. That leads to the inevitable conclusion that whatever shape you design, if built on a large enough scale, will collapse under its own weight. An elephant-sized beetle would never get off the ground; its legs could not sustain the weight.

To return to the Statue of Liberty: if you double the size of the statue, then you double the length of every curve on the statue, you quadruple the amount of paint required to cover the surface of the statue, and you need eight times as much material in the construction. In general, if you increase the size of an object by any given factor, then the length of every curve on it gets multiplied by the same factor, the area of any surface is increased by the *square* of the factor, and the volume of the object is multiplied by the *cube* of the factor.

This relationship to scaling helps to sum up the very essence of dimension. A curve is *one*-dimensional in that its length changes by exactly the scaling factor. (A circle of diameter 1 has length π, and a circle of diameter 3 has length 3π.) A surface is *two*-dimensional because its area goes up by the *square* of the scaling factor. (The area inside a circle of radius 1 is π; the area inside a circle of radius 3 is 9π.) A solid is *three*-dimensional because its volume is multiplied by the *cube* of the scaling factor. Another way to look at it is that the *dimension* is simply the *exponent of the scaling factor* needed to determine the corresponding magnitude in the scaled-up object.

For the snowflake curve, our measuring stick had the peculiar property that when you scaled it up by a factor of three, it increased in "size" by a factor of four, since it ended up consisting of exactly four copies of the original measuring stick. If it were truly a "curve," then its size should go up by the same factor of three. If it were two-dimensional, then its size should go up by the *square* of three, or nine. Because of this, Hausdorff decided that the snowflake curve's dimension is *greater* than one, but *less* than two. In fact, he gives an exact value for the dimension, which turns out to be a little over one and a quarter.

All of this may seem far removed from everyday reality—even the reality of an actual snowflake. But it turns out to be closely tied to one of the fundamental issues in cosmology: the "cosmological principle" itself—the principle that addresses the global distribution of galaxies in the universe. According to the cosmological principle, when viewed on a large enough scale, the distribution of mass in the universe is like homogenized milk: no cream in one part and watery liquid in another, but the same wherever you look. It is clear that that is *not* the case when the universe is viewed on smaller scales. There *are* stars and clusters of stars, galaxies and clusters of galaxies, and superclusters of galaxy clusters.

The astronomer and cosmologist Gérard de Vaucouleurs suggested in a famous article published in 1970 entitled "The Case for a Hierarchical Cosmology" that there was no more reason to assume that the distribution of galaxies would eventually smooth out in "homogenized" fashion than to conclude that the clustering of stars and galaxies continues throughout the universe, regardless of scale. On the basis of observational data available at the time, de Vaucouleurs calculated the distribution of galaxies in a numerical fashion that Mandelbrot later interpreted to mean that the geometry of all the galaxies in the universe represents a fractal whose dimension is just under one and a quarter (a little less than that of the snowflake curve).

During the quarter century since de Vaucouleurs wrote his article, larger and larger sky surveys have been undertaken, partly with the goal of trying to settle the question of large-scale smoothing versus clustering. In 1985, Margaret Geller, John Huchra, and Valerie de Lapparent at Harvard University's Smithsonian Astrophysical Observatory completed a different kind of survey, in which they chose a long, thin strip of the sky and made a survey that focused on the depth dimension—how far away each galaxy in that strip was from us. The most striking result of this more three-dimensional sky survey was the detection of what appeared to be a new large-scale structure in the universe: a kind of soapsuds or froth effect, with large "bubbles" containing very few galaxies in their interiors, surrounded by a thin "skin" in which many galaxies were clustered. In subsequent surveys of adjacent strips of the sky, the same team produced more maps that tended to confirm their original finding. Combining the strips to produce a three-dimensional picture of a substantial slice of the universe revealed quite clearly a concentration of galaxies on a much larger scale than any previously detected. That concentration, stretching along almost the entire width of their map, has been dubbed "the great wall," and has led to further doubts about the smoothness of the universe on a sufficiently large scale. As a result, a fractal universe is still not out of the question. On the other hand, attempts to reconcile a fractal structure for the universe with all the observational data, including the remarkably smooth cosmic microwave background radiation, have not been particularly successful.

Of course, there are many possibilities regarding the nature of the universe besides smooth and fractal. What is truly exciting is that observational techniques have advanced so rapidly that mathematicians and cosmologists may soon have definitive answers to these and other fundamental questions about the shape and texture of the universe.

But even as that is happening, geometers continue to create new geometries, each a fascinating world of its own, inhabited by previously unsuspected species, where even notions of "shape" and "size" may take on new meaning. The lesson of the past is that we cannot tell in advance how long we may have to wait to find a use for each new mathematical creation or where in the real world it might arise. In the meantime, we may picture the product of three thousand years of geometric inventiveness in the form of a tree—the "Geometree"—whose roots go back even further and whose branches represent the outcome of centuries of discovery and creation. With or without applications, the branches and fruits of this tree are worth contemplating as a remarkable product of the human imagination. As we approach the end of another millennium, the "Geometree" is healthy, vigorous, and in full foliage, older than any redwood, and fully as majestic.

Postlude

Richard Feynman, the brilliant and idiosyncratic American scientist, was a close observer and connoisseur of the intricate dance that those partners—physics and mathematics—seem to perform endlessly: now tightly clasped together, almost indistinguishable, now spinning farther and farther apart, eyeing each other warily at a distance. In *The Character of Physical Law,* Feynman writes, "Every one of our laws is a purely mathematical statement in rather complex and abstruse mathematics . . . Why? I have not the slightest idea." And later, "To those who do not know mathematics it is difficult to get across a real feeling as to the beauty, the deepest beauty, of nature."

I have tried in this book to trace a path through the mathematical landscape that leads to a view of one aspect of nature—the nature of the cosmos. By the *cosmos,* I mean the universe in its entirety, possessing an order, structure, and shape on its largest scale. That shape is not discernible or even describable without the language of mathematics. Studying mathematics in order to understand the laws of physics is not unlike learning enough of a foreign language to capture some of the special flavor and beauty of prose or poetry written in that language. In the process, one may well become fascinated by the language itself. And so it is with many parts of mathematics. Created in the first instance to provide deeper insights into the nature of the world about us, the language of mathematics develops its own structure and order, its own beauty and fascination. I hope that the dual nature of mathematics—its internal beauty and its power to reveal the hidden structure of the external world—will have become apparent in the course of this narrative.

Acknowledgments

My primary debt is to James L. Adams, who confronted me many years ago with the query: How is it that mathematics is such a beautiful subject, yet students can go through four years of college taking many math courses and never find out? He and I embarked on a project to correct that situation. We were joined by Sandy Fetter in designing and teaching a course at Stanford University called "The Nature of Science, Mathematics, and Technology" in which we tried to convey the essence of each of our subjects and the multiple ways in which they were linked together. This book had its genesis in one segment of the course, devoted to astronomy and cosmology, where we showed how the technology of astronomical instruments, the science of astrophysics, the theory of relativity, and the mathematical subjects of geometry and topology all combined to produce the remarkable picture we now have of the entire universe. I am further indebted to Jim and Marian Adams for their continued support and hospitality during the long evolution from the early planning stages, through the course itself, and to the present book.

I am also indebted to a number of individuals who

played a significant role during various stages in the writing of the book:

Wilbur Knorr for being always available to answer questions on the history of mathematics and astronomy, especially concerning the ancient Greeks.

Michael Barall for many enlightening conversations about cosmology, and for numerous suggestions, corrections, and comments on early versions of the manuscript.

Paul Alpers, Gordon Godberson, David Hoffman, and Henry Landau for reading the entire manuscript, for many valuable comments, and for assistance in a variety of ways.

Jo Butterworth for many forms of help, particularly in her role as librarian; also Ralph Moon for additional library assistance, and the University of California at Berkeley library system for the happy combination of mathematics and astronomy in one library.

Elizabeth Katznelson and Sharlene Pereira for patiently and expertly transforming each new stage of densely annotated manuscript into clean and accurate typescript.

Susan Bassein for going to extraordinary lengths to produce accurate and attractive illustrations.

Jill Kneerim for being far more than an excellent agent, offering sage advice and help through the entire process of publishing the book.

Roger Scholl, my editor at Anchor, for wrestling tirelessly with every detail of the writing, preparation, and production of the book, and for a number of suggestions that have been adopted in the final version.

Among the many others who have assisted in one way or another with the long process of making this book a reality, I would like to thank Wendy Lesser, Gayle Greene, Diane Middlebrook, Carl Djerassi, Barrett O'Neill, Alexander Fetter, Hermann Karcher, Robert Jourdain, Nancy Shaw, Bud Squier, Judy Squier, Freda Birnbaum, Arlene Baxter, Julie Driscoll, William Blackwell, Joe Christy, Jean-Pierre Bourguignon, and numerous colleagues and staff members at both Stanford University and the Mathematical Sciences Research Institute in Berkeley, as well as the many students whose astute questions and expressions of interest or puzzlement have served as a guide and inspiration in my writing.

Finally, I am greatly indebted to the Sloan Foundation for their generous financial support during the preparation of both the course that led to this book and the book itself.

NOTES

I am indebted to the following books as sources for many of the quotations used in this book:

Cole, K. C. *Sympathetic Vibrations: Reflections on Physics as a Way of Life.* New York: William Morrow, 1985 (Feynman quotation on p. 220).

Moritz, Robert Edouard. *Memorabilia Mathematica: The Philomath's Quotation Book.* New York: Macmillan, 1914. Reprinted by the Mathematical Association of America, 1993.

Schmalz, Rosemary. *Out of the Mouths of Mathematicians: A Quotation Book for Philomaths.* Washington, D.C.: Mathematical Association of America, 1993.

The Dyson quotation in Chapter 8 is from "Mathematics in the Physical Sciences," *Scientific American,* September 1964.

Preface

p. xi "newspapers . . . reported": The announcement of the discovery was made by George Smoot of Berkeley, who headed the team carrying out the research.

p. xi "snapshot": The picture presented is a "snapshot" in the sense of capturing a particular moment in the history of the universe. It is far from a "snapshot" in the way it was produced; it is the end result of processing vast quantities of raw data using elaborate statistical methods and solving a set of approximately 6,000 mathematical equations, then translating the results into pictorial form.

p. xi "the moment that space began": The technical name for that phase in the evolution of the universe is *decoupling.* To say that "space began" at that moment means that "space" in the sense of space between objects, or of "outer space," did not exist prior to the time of decoupling.

There were still spatial measurements, just as there are in the interior of the sun.

Chapter 1

p. 4 "the Babylonians": See Otto Neugebauer's *The Exact Sciences in Antiquity,* Brown University Press, Providence, 1957.

p. 5 "new discoveries": Some of the earliest surviving examples of geometric proofs date from the 5th century B.C. One example, due to Hippocrates of Chios, is the surprising fact that one can give an exact value for the area of the shaded crescent-shaped region in the figure below. Hippocrates gave a complete proof that its area is exactly the same as the area of the triangle shown in the figure.

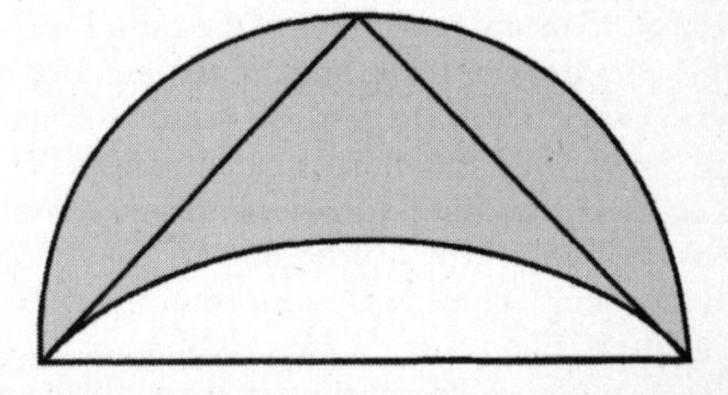

For more background, see Wilbur Knorr's book *The Ancient Tradition of Geometric Problems,* Dover, New York, 1993, pp. 26–32.

p. 5 "Euclid's life . . . Pythagoras": Much of the information cited here can be found in the *Oxford Classical Dictionary.*

p. 7 "for better and for worse": The fascination with circles amounted in some cases to an obsession. Copernicus as well as Ptolemy built a solar system by piling circles upon circles, and even Kepler was long blinded to the evidence he had accumulated before realizing that planetary orbits are not circles, but ellipses. (See the excellent recounting of that story in Arthur Koestler's *The Watershed,* Anchor, New York, 1960; reprinted by the University Press of America, 1984. "The Watershed" also appears as Part IV of Koestler's *The Sleepwalkers,* Grosset & Dunlap, New York, 1959.)

p. 8 "noted by Aristotle": From Aristotle's book *On the Heavens,* Book 2, Chapter 14.

p. 9 "higher latitudes": The *latitude* of a point on the earth is a measure of the distance from the equator. It is measured in *degrees,* starting with zero at the equator and reaching 90 degrees at the poles. The *parallels of latitude* are the circles, parallel to the equator, at a given latitude.

p. 11 *"winter solstice":* Two other key days of the year are the spring and autumn equinoxes. On those days the sun rises exactly in the east and sets in the west. On all other days it rises either south (in the winter) or north (in the summer) of east, and the tip of the gnomon's shadow traces out a path bending northward in the winter and southward in the summer. On the equinoxes, the tip of the shadow describes a straight line in the east-west direction.

p. 13 "the angle between the vertical directions at Alexandria and Aswan": A key point in Eratosthenes' calculation is that the distance to the sun is so great compared with all the other distances involved that to any conceivable accuracy of measurement at that time the direction of the sun as seen from any two points of the earth may be considered to be the same; in other words, the lines from Alexandria and Aswan to the sun may be assumed to be parallel. What differs is the "vertical" direction at each point, and that difference precisely reflects the curvature of the earth. The pictures accompanying an account of Eratosthenes' method in some widely read books *(The Story of Maps,* by Lloyd A. Brown, Brown and Company, Boston, 1949, p. 31; *The Shape of the World* by Simon Berthon and Andrew Robinson, Rand McNally, Chicago, 1991, p. 23) show the direction of the sun to be markedly different at the two locations on earth, making it appear that it is the angle between those two directions that enters in the calculation. In fact, the angle between the directions toward the sun from Alexandria and Aswan is well under a thousandth of a degree.

p. 14 "Eratosthenes' method": For a modern discussion of Eratosthenes' goals and procedures, along with references to sources, see Bernard R. Goldstein, "Eratosthenes and the 'Measurement' of the Earth," *Historia Mathematica* 11 (1984), pp. 411–16.

p. 15 "they all look alike": Another basic property of circles is that one can determine the whole course of a circle from any small arc on it. That property is what allowed Eratosthenes to deduce the full circumference of the earth from the small arc between Alexandria and Aswan. The same property was used by Proust in *Remembrance of Things Past* to illustrate the mistaken notion that somebody should be interested in meeting another person "just like you."

p. 16 "mentioned in the Bible": I Kings 7:23: "And he made a molten sea, ten cubits from one brim to the other: it was round all about . . . and a line of thirty cubits did compass it round about."

This passage has sometimes been interpreted to mean that according to the Bible, the value of π is exactly equal to 3. However, such an interpretation ignores the fact that the standard practice, then as now, is to round off measurements to whatever degree of accuracy is relevant for a given purpose.

p. 17 "al-Kashi": My primary source for mathematics and astronomy in the Islamic period has been J. L. Berggren, *Episodes in the Mathematics of Medieval Islam,* Springer, New York, 1986. Also useful was the classic by J. L. E. Dreyer, *A History of Astronomy from Thales to Kepler,* Dover, New York, 1953.

p. 17 "to sixteen decimal places": Whatever al-Kashi's stated motive may have been, the urge to calculate π to greater and greater accuracy seems to transcend time and geography. Here are some of the records for the number of digits of accuracy surpassed:

five	5th century	Tsu Ch'ung-chih	China
ten	1424	al-Kashi	Samarkand
one hundred	1706	John Machin	London
one thousand	1949	J. W. Wrench	U.S.
ten thousand	1957–58	G. E. Felton	London
		F. Genuys	Paris
one hundred thousand	1961	Daniel Shanks & J. W. Wrench	Washington, D.C.
one million	1974	J. Guilloud & M. Bouyer	Paris
ten million	1985	Kanada	Tokyo
one hundred million	1987	Kanada	Tokyo
one billion	1989	David & Gregory Chudnowsky	New York
		Kanada	Tokyo

An article by Richard Preston in the April 1992 issue of *The New Yorker,* entitled "Mountains of π," describes the life and work of the Chudnowsky brothers and their record calculation of over two billion digits of π as of that time.

A highly readable, if idiosyncratic, account of many related matters is *A History of π* by Petr Beckman, St. Martin's Press, New York, 1974.

p. 17 "Tsu Ch'ung-chih": Also written Zŭ Chōngzhī, 429–500 A.D. For more of his accomplishments and those of other Chinese mathematicians, see Lĭ Yan and Dŭ Shíràn, *Chinese Mathematics: A Concise History.* Also Joseph Needham, *Science and Civilization in China.*

Chapter II

p. 20 "Christopher Columbus": For the voyages of Columbus and other voyages during the age of exploration, my main reference has been Samuel Eliot Morison, *Admiral of the Ocean Sea: A Life of Christopher Columbus,* Little, Brown, Boston, 1942. Among the many popular books on the subject are: Björn Landström, *Columbus,* Macmillan, New York, 1967; Daniel J. Boorstin, *The Discoverers,* Vintage Books, New York, 1985; Lloyd A. Brown, *The Story of Maps,* Little, Brown, Boston, 1949; and Simon Berthon and Andrew Robinson, *The Shape of the World,* Rand McNally, Chicago, 1991. Some specific references to these books are made in the notes below, where they are cited by the authors' names.

p. 21 *"Almagest":* Ptolemy's own title for the *Almagest* was *Mathematical Syntaxis.* See O. Pedersen, *A Survey of the Almagest,* Odense, Denmark, University Press, 1974.

p. 22 "Sacrobosco": My main source for Sacrobosco and *The Sphere* was Lynn Thorndike, *The Sphere of Sacrobosco and Its Commentators,* University of Chicago Press, Chicago, 1949.

p. 23 "the stars would rise as soon for westerners as for orientals": Although he does not elaborate, it seems most probable that the way a given moment would be singled out was by a celestial event, such as a lunar eclipse, which would happen simultaneously at different locations on earth. Noting the position of various stars in the sky at such a moment would have led to the observation of different rising times referred to by Ptolemy. In fact, it may well be that theory preceded observation in this case. Ptolemy's argument that the earth is curved from north to south, because stars visible at one latitude cannot be seen at another farther north, is cited hundreds of years earlier by Aristotle *(On the Heavens,* Book 2, Chapter 14, where a number of additional reasons are given for the earth to be spherical in shape). Since the sphere of the stars was pictured as rotating around the spherical earth, the conclusion was inescapable that the moment a given star first appeared on the horizon would depend on the location from east to west.

Incidentally, it was precisely by carefully noting the local time of a lunar eclipse that Columbus was able to establish for the first time the longitude of the islands he had discovered in the Caribbean. See Morison, p. 655.

p. 24 "the workings of gravity": This objection to Columbus' proposal is related by Columbus' son Fernando. See Landström, p. 39.

p. 25 "Junta dos Matemáticos": See Morison, p. 69, for the composition of the Junta.

p. 29 "compass directions be correctly depicted on the map": What is meant specifically by the second condition is that angles on the map are equal to the corresponding angles on the surface of the earth. That property of a map is called *conformal.* For example, if two roads intersect at a certain angle, then the lines representing the roads on the map should intersect at the same angle. Combining this property with the first one—that a north-south line on the earth be depicted by a vertical line on the map—results in east-west lines being depicted as horizontal and all compass directions translating similarly on the map.

p. 30 "a fixed scale along each parallel of latitude": One way to see why that is true is the following. Since north-south corresponds to the vertical direction on the map, it follows that each meridian on earth will correspond to a vertical line on the map. Since compass directions are correctly depicted, the east-west direction at any point corresponds to the horizontal direction on the map. Hence each parallel of latitude on the earth corresponds to a horizontal line on the map. The distance between two such nearby horizontal lines on the map, divided by the distance along the earth's surface between the corresponding parallels of latitude gives the scale of the map in the vertical direction, which must therefore be the same at every point along a given parallel of latitude. (Strictly speaking, one must take the limit of this ratio, as the lines get closer together.) But it is a well-known fact from linear algebra that if directions are preserved at a point, then the scale in the horizontal direction must equal the scale in the vertical direction at the point. Hence the scale in the horizontal direction is the same at every point of any given parallel of latitude.

p. 32 "the principles on which the map was drawn": The idea behind Mercator's map is this: If we want the north-south direction on earth to correspond to the vertical direction on our map, then each meridian must be depicted by a vertical line on the map. Since two meridians are farthest apart where they cross the equator and get closer and closer together as they approach the poles, while the vertical lines representing them on the map are parallel and remain a fixed distance apart, it follows that east-west directions between meridians are stretched more and more on the map as we move from the equator toward the poles. The key idea that Mercator had was that in order to make directions come out right, the map had to stretch distances *along* the meridians by exactly the same amount that they were being stretched *between* meridians; that is, the amount of horizontal and vertical stretching at each point should be the same. What Mercator did was to draw a map conforming as closely as possible to that property.

For more on the story of the mathematics behind Mercator's map, see the article by F.V. Rickey and Philip M. Tuchinsky in *Mathematics Magazine*, Vol. 53 (1980), pp. 162–66.

p. 33 "the exact equations for Mercator's map": If we use standard rectangular coordinates to define the map, with a horizontal x-axis and vertical y-axis, then we can place the equator along the x-axis. The x-coordinate of any point will be a fixed multiple of its longitude while its y-coordinate will be a multiple of the expression

$$\log \frac{1 + \sin d}{\cos d}$$

where d is the number of degrees of latitude (taken as positive north of the equator and negative south of the equator). At the poles, this expression becomes infinite, which is responsible for two of the chief objections to Mercator's map: first, it must always leave out a region about each of the poles, and second, sizes are progressively more exaggerated as the distance from the equator increases. However, the simple cylindrical projection described in the text suffers from both of these flaws, and to even greater degree.

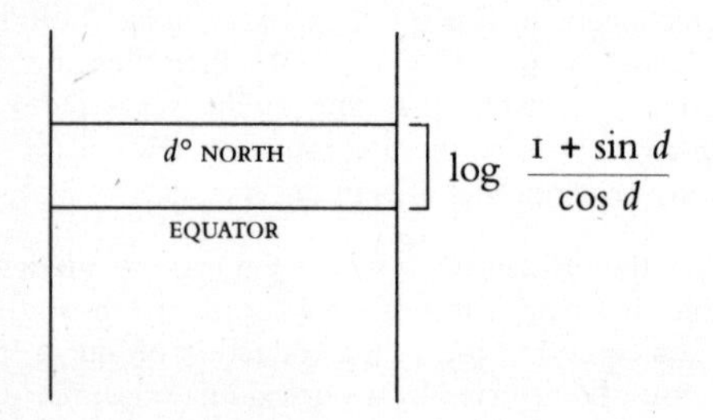

p. 33 "the use of logarithms": Logarithms were invented by John Napier in 1614.

p. 34 "integral calculus": The integral calculus is a mathematical technique for evaluating the cumulative effect of a continually changing quantity. For example, if one could have a map with a fixed scale, then one could obtain actual distances on the earth from measured distances on the map by simply multiplying by the scaling factor (which is what is done for small regions where the scaling factor is almost the same everywhere). In the case of Mercator's map, the scaling factor varies with the latitude. Moving due north on the earth corresponds to moving vertically on the map with a continually changing scale. The cumulative increase in height on the map is obtained as the "integral" of the scaling factor and results in the expression given above.

p. 34 "a geometrical fact": The proof that every navigator's map is a Mercator map depends on the property we noted earlier for every navigator's map: the scale is fixed along horizontal lines. Furthermore, as we saw in a previous note, the exact value of that scale is easily determined for each latitude. Then, for angles to be preserved, the north-south scale must match the east-west scale at each latitude. That means we know the rate of change of the vertical coordinate on the map with respect to the latitude on the earth, which turned out to be a constant multiple of $1/\cos d$. The explicit formula is then obtained by integration, as we saw in the case of Mercator's map.

p. 36 "Every map, in fact, is a compromise": Euler's proof that there is no exact scale map of any portion of a sphere dates from 1775, when he presented it to the St. Petersburg Academy of Science in a paper entitled "On Representations of a Spherical Surface on the Plane." The paper is in Latin and is reprinted in Euler's collected works, Series 1, Vol. 28, pp. 248–75. Euler's proof, on pp. 251–53 of his paper, is a classical "proof by contradiction": he assumes that there *is* an exact scale map, and by making a number of computations based on that assumption, arrives at a contradiction; it follows that the original assumption—that a scale map exists—must be false.

p. 37 "several dozen are in common use": The United States Geological Survey, for example, uses sixteen different map projections for its published maps. See John P. Snyder, *Map Projections Used by the U.S. Geological Survey,* Geological Survey Bulletin 1532, U.S. Government Printing Office, Washington, D.C., 1983, for details on the maps used as well as historical background. A more extensive collection of some hundred different types of maps, with pictures and equations for each, can be found in *An Album of Map Projections* by John P. Snyder and Philip M. Voxland, U.S. Geological Survey Professional Paper 1453, U.S. Government Printing Office, Washington, D.C., 1989.

p. 37 *"azimuthal equidistant projection":* See J. L. Berggren, *Episodes in the Mathematics of Medieval Islam,* Springer, New York, 1986, p. 10, and for more detail, Berggren's "Al-Biruni on Plane Maps of the Sphere" in *Journal for the History of Arabic Science,* Vol. 6, 1982, pp. 47–80; esp. p. 67.

Chapter III

p. 41 "Gauss and . . . Beethoven": Beethoven was born on December 16, 1770, in Bonn; Gauss on April 30, 1777, in Brunswick. The parallels between Gauss and Beethoven (developed somewhat further in Chapter 5) include their physical descriptions as strongly built, short, and stocky. Also, they were both born into poor households and were subject to the whims of harsh, abusive fathers.

p. 41 *"princeps mathematicorum":* This Latin expression has also led to Gauss's frequent appellation: "Prince of Mathematicians." It is common among mathematicians to place only two mathematicians in history at the same level as Gauss: Archimedes and Newton.

p. 42 "the telegraph": For more on the history of the telegraph and the role of Gauss and his collaborator Weber, see p. 129 of W. K. Bühler's *Gauss: A Biographical Study,* Springer, New York, 1981.

p. 43 "the 'error curve' ": The equation for this curve is $y = e^{-x^2}$. Another common name for it is the "normal distribution."

p. 44 "all the numbers from one to a hundred": Different versions of this story have different sets of numbers, but all are based on the same principle.

p. 46 "a whole class of more general problems": The general problem is to find the sum of an *arithmetical series,* a series in which the difference between two successive terms is a fixed number *d*.

p. 47 "bulge at the equator": A cross section of the earth through the poles is approximately an ellipse whose shorter axis—the distance from the North Pole to the South Pole—is 7,900 miles, and whose longer axis—the earth's diameter at the equator—is 7,926 miles.

p. 51 "degrees of longitude": Since the circumference of the earth at the equator is a little short of 25,000 miles, and there are 360° in a full circle, it follows that the length of a one-degree arc at the equator is somewhat less than 25,000 divided by 360, or a bit under 70 miles.

p. 52 *"Gauss curvature":* The study of various notions of curvature is a central part of the branch of mathematics known as *differential geometry.* For a curve, the curvature is defined at each point as the reciprocal of the radius of the circle that best approximates the curve near that point. For a surface, one has at each point two *principal curvatures* defined as the maximum and minimum values of the curvature at the given point of the curves obtained by slicing the surface by a plane through the point perpendicular to the surface. (The curvature of those curves is also given a sign—positive or negative, depending on which side of the plane tangent to the surface the curve lies on.) The Gauss curvature is the product of the principal curvatures. If it is positive at a point, then the surface near the point lies on one side of the tangent plane at the point, as for a sphere or ellipsoid. If it is negative, then the surface crosses the tangent plane, as in the case of a surface shaped like an hourglass at a point on the waist.

p. 54 *"geodesic":* For most people the word "geodesic" evokes the name Buckminster Fuller and his well-known dome.

For a fascinating account of geodesic domes, see "Geodesics, Domes, and Spacetime," Chapter 3 of *Science à la Mode,* by Tony Rothman, Princeton University Press, Princeton, 1989.

p. 56 "more complicated formulas": If s is the sum of the angles of a geodesic triangle on a sphere of radius r (so that each side is an arc of a great circle) and A is the area of the triangle, then the formula for A is

$$A = \pi r^2 \left(\frac{s}{180} - 1\right).$$

(This formula was first published by the Flemish mathematician Albert Girard in 1629.) By measuring the area A and the sum of the angles s, this equation can be used to determine the radius r of the sphere.

p. 56 "The gist of the formula": The precise formula given by Gauss involves the notion of a surface integral. If s is the sum of the angles, measured in degrees, for a geodesic triangle on a surface, then

$$s = 180 + \frac{180}{\pi} \int K dA$$

where K is the Gauss curvature and the integral is taken over the interior of the triangle. In the plane, $K = 0$ and we have $s = 180$ for any triangle. On a sphere of radius r, we have $K = 1/r^2$ and $\int K dA = A/r^2$, where A is the area of the triangle. Then $s = 180(1 + A/\pi r^2)$, which is the same as Girard's equation given above for a spherical triangle. In general, on a surface with positive curvature we have $K > 0$, $\int K dA > 0$, and the sum of the angles s is greater than 180 degrees. Negative curvature implies $K < 0$ and s is less than 180 degrees.

p. 57 "just by making measurements on the surface": For practical purposes one would not want to compute curvature by just using measurements on a surface if one has other means available, since it requires very accurate and often difficult measurements, such as the area of a triangle or the circumference of a circle. However, the fact that it is possible in theory has important consequences: the impossibility of making exact scale maps. Also, as we shall see when we get to Riemann, there may be cases where there are no other means available.

p. 57 "a 'circle' . . . on a . . . surface": The technical name for such a "circle" is a *geodesic circle.*

p. 59 "the same is true on any surface": The formulas of Bertrand and Puiseux give the following expression for the Gauss curvature at a point p on a surface. Let $L(r)$ be the length of the circle of radius r on the surface with center at p. Then the expression

$$\frac{3}{\pi}\,\frac{2\pi r - L(r)}{r^3}$$

is very close to the value of the curvature of the surface at the point p when r is small. The exact value of the curvature is given by a limit:

$$K = \lim_{r \to 0} \frac{3}{\pi}\,\frac{2\pi r - L(r)}{r^3}.$$

If the curvature is positive, $K > 0$, then the circumference $L(r)$ falls short of the euclidean value $2\pi r$, while $K < 0$ means that $L(r)$ is greater than $2\pi r$. A good approximation to $L(r)$ is $2\pi r - K\pi r^3/3$.

p. 60 "a paper written in 1827": Gauss's paper was entitled *"Disquisitiones Generales Circa Superficies Curvas,"* or "General Investigations of Curved Surfaces." Both the original and an English translation are reprinted along with an excellent survey of work leading up to it and subsequent developments in Peter Dombrowski, "150 Years after Gauss' *Disquisitiones Generales Circa Superficies Curvas," Astérisque* 62, Société Mathématique de France, Paris, 1979.

Chapter IV

p. 62 "imaginary number": The *imaginary numbers* are all multiples of i by real numbers: $2i$, $\sqrt{3}i$, $-i$, etc. *Complex numbers* are sums of real and imaginary numbers, such as $2 + 3i$.

p. 63 "non-euclidean geometry": An excellent source for all matters related to non-euclidean geometry is the book by B. A. Rosenfeld, *A*

History of Non-Euclidean Geometry: Evolution of the Concept of a Geometric Space, Springer, New York, 1988.

p. 63 "Immanuel Kant": In his *Critique of Pure Reason,* where he characterizes euclidean geometry as a priori knowledge: part of our perception of the world, rather than based on experience. For a more detailed discussion of Kant's views on euclidean geometry, see Chapter 3 of Richard J. Trudeau's *The Non-Euclidean Revolution,* Birkhäuser, Boston, 1987, and Part One of Michael Friedman's *Kant and the Exact Sciences,* Harvard University Press, Cambridge, 1992.

p. 63 "Nikolai Ivanovich Lobachevsky": Some readers may be familiar with Tom Lehrer's song "Lobachevsky," with its ringing refrain: "Plagiarize!," ascribed to Lobachevsky. If so, they may wonder what Lobachevsky may have done to deserve that. The answer is "nothing." His name just happened to fit the needs of the lyrics.

p. 63 "sum of the angles is . . . 180°": From the axiomatic point of view, what distinguishes euclidean from non-euclidean geometry is the status of Euclid's fifth postulate. In a form different from but equivalent to the one used by Euclid, the fifth postulate asserts the existence of a unique line parallel to a given line, through a given point not on the line. That property turns out also to be equivalent to the condition that the sum of the angles in a triangle is 180°. Lobachevsky replaces Euclid's fifth with the postulate that there is *more* than one line parallel to a given line through a point not on the line; that in turn is equivalent to the property that the sum of the angles in a triangle be *less* than 180°.

p. 64 "Bolyai's father, Wolfgang": Wolfgang was the German name that János Bolyai's father used in certain contexts. His Hungarian name is Farkas, which one often sees used.

p. 64 "a less admirable side": This aspect of his character emerged in another case of a brilliant young mathematician showing an epoch-making discovery to Gauss, only to be summarily dismissed. In that case it was Niels Henrik Abel, a twenty-three-year-old Norwegian, who had settled in a most surprising fashion one of the most fundamental questions in algebra, concerning solutions of algebraic equations. It had been known since antiquity that every quadratic equation, such as $x^2 - 3x + 4 = 0$, could be solved by an explicit formula. Cubic equations, such as $x^3 - 3x^2 + 5x - 2 = 0$, turned out to be much harder, and it was not until the sixteenth century that methods were found to solve a general cubic equation; that is to say, to find a way of expressing a solution in terms of the coefficients. (The expression for a general quadratic equation: $ax^2 + bx + c = 0$, is $x = (-b \pm \sqrt{b^2 - 4ac})/2a$). A general fourth-degree equation, starting with x^4, and then lower powers, can also be solved explicitly, but 300 years of effort with *fifth*-degree equations led nowhere. Abel first thought he had found a solution, but after he saw a mistake in what he had done, he arrived at an astonishing result: he proved that a general solution was *impossible.* It was this masterpiece that Gauss apparently declined to look at closely enough to recognize its value.

p. 65 "Gauss had in fact anticipated": Although Gauss never published anything on the subject, the extent of his investigations is quite clear from a number of letters that have been preserved.

p. 66 "question in geodesy": See "The Myth of Gauss' Experiment on the Euclidean Nature of Physical Space" by Arthur I. Miller, *Isis* 63 (1972), pp. 345–48.

p. 66 " 'Imaginary Geometry' ": The exact reference is N. Lobatschewsky (one of the many variants in the spelling of his name), Géométrie imaginaire, *Crelle's Journal*, Vol. 17 (1837), pp. 295–320.

p. 67 *"Crelle's Journal":* The full name of the journal is *Crelle's Journal der Mathematik.* August Leopold Crelle (1780–1855) was a remarkable and perhaps unique figure in the annals of mathematics. A civil engineer by trade, he was a mathematical enthusiast with the energy and know-how to influence the course of events. In his professional life, Crelle was in charge of building the first railroad in Germany. Starting in 1828, he was a consultant in mathematics to the Prussian Ministry of Education, and in that capacity he was able to exert considerable influence on the way mathematics was perceived and presented in school. He believed that it was a mistake to view mathematics mainly from the perspective of applications—that considerations internal to mathematics should be primary both in the way mathematics evolves and in the way that it is presented.

In 1826, he founded his new journal devoted exclusively to mathematics. He started it off in grand form by publishing a whole series of papers by Abel, including the great one that proved the unsolvability of the general fifth-degree equation.

p. 67 "triangles on a pseudosphere": Actually, Minding does not single out the pseudosphere, but states that the same equations hold for geodesic triangles on any surface of constant negative curvature. He refers to the paper that he published the previous year for explicit examples of surfaces of constant negative curvature, including the pseudosphere.

p. 67 "one such formula": On p. 324 of Minding's paper in *Crelle's Journal,* Vol. 20 (1840), pp. 323–27. The formula given by Minding is

$$\cos a\sqrt{k} = \cos b\sqrt{k} \cos c\sqrt{k} + \sin b\sqrt{k} \sin c\sqrt{k} \cos A$$

where a, b, c are the lengths of the sides of a triangle on a surface of constant curvature k, and A is the angle opposite the side of length a. If the surface is a sphere of radius R^2, then $k = 1/R^2$, and the resulting formula was well known for a triangle on a sphere. If the surface has negative curvature $k = -1$, then $\sqrt{k} = \sqrt{-1} = i$, and using the expressions $\cos ix = \cosh x$, $\sin ix = i \sinh x$, where $\cosh x$ and $\sinh x$ are the so-called hyperbolic cosine and sine of x, one gets the equation

$$\cosh a = \cosh b \cosh c - \sinh b \sinh c \cos A$$

which translates to equation (10) on p. 298 of Lobachevsky's 1837 paper in *Crelle's Journal,* after substituting the notation given on the top of p. 296. In this case, a, b, c are the lengths of sides of a triangle on a surface with constant Gauss curvature $K = -1$, and A is again the angle opposite side a. This equation is the non-euclidean analog of a formula familiar to students of trigonometry:

$$a^3 = b^2 + c^2 - 2bc \cos A.$$

p. 68 "leading German mathematician of his time": Lambert was in fact the *only* famous German mathematician of his time. He died in 1777, the year that Gauss was born, leaving Germany without a single mathematician of repute until Gauss himself became active. The astonishing transformation of Germany from a mathematical desert in the late eighteenth century to a world center in the mid-nineteenth is the theme of a recent book: *Möbius and His Band,* edited by John Fauvel, Raymond Flood, and Robin Wilson, Oxford University Press, Oxford, 1993.

p. 68 "π is *irrational*": For a remarkably short modern proof that π is irrational, see Ivan Niven, *Irrational Numbers,* Carus Mathematical Monographs, No. 11, Mathematical Association of America, 1956, pp. 19–21.

p. 75 "necessarily distorts lengths and distances": The reason that distortion is inevitable in any model representing the Lobachevsky plane mapped onto a plane domain is Gauss's theorem that exact maps preserve curvature. Since Lobachevsky's geometry corresponds to geometry on a surface of negative curvature, any exact-scale map would also have negative curvature. But the plane has curvature zero, and an exact-scale map of the plane is therefore impossible.

Chapter V

p. 79 " 'law of inertia' ": The idea of casting Galileo's law of inertia in the light of a thought experiment is taken from the book *The Evolution of Physics* by Albert Einstein and Leopold Infeld, Simon & Schuster, New York, Touchstone edition, pp. 5–11. Incidentally, the Einstein-Infeld book, subtitled *From Early Concepts to Relativity and Quanta,* written in 1938, remains a masterpiece of exposition, explaining fundamental ideas of physics in a nontechnical fashion.

p. 82 "*define* the curvature of space": If $L(r)$ is the circumference of the circle of radius r in the equatorial plane, then the "curvature of space in the equatorial plane of the earth" is approximately

$$\frac{3}{\pi}\,\frac{2\pi r - L(r)}{r^3}.$$

This expression should depend very little on the choice of radius r. By choosing a sequence of circles, we can see if we are in the range where the value does not depend significantly on r. The precise theoretical value of the curvature of space in the equatorial plane is given as a limiting value:

$$\lim_{r\to 0}\frac{3}{\pi}\,\frac{2\pi r - L(r)}{r^3},$$

precisely the same value, according to the formula of Bertrand and Puiseux, as the curvature of a surface at a point where $L(r)$ is the length of a circle of radius r centered at that point.

p. 83 "If space is euclidean": In ordinary (euclidean) geometry, six points equally spaced on a circle centered at a given point form a regular hexagon, composed of six equilateral triangles; the distance between adjacent points is equal to the distance of each point to the center.

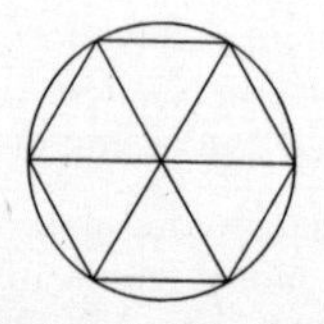

p. 87 " 'equatorial sphere' ": The reason for this terminology is that if we think of the earth divided into northern and southern hemispheres, rather than eastern and western, then the two boundary circles represent the same great circle in the sphere—the equator—seen from one side by residents of the northern hemisphere and from the other side by residents of the southern hemisphere.

p. 89 "the *Divine Comedy*": The relevant passage is from Canto 28 of *Paradiso,* lines 1–129.

p. 89 "a center where a point of light radiates": In the translation by Charles S. Singleton, Princeton University Press, Princeton, 1975, Dante writes: "I saw a point which radiated a light so keen that the eye on which it blazes needs must close . . ."

p. 90 "the universe according to Dante would coincide": The fact that Dante's depiction of the universe in the *Divine Comedy* could be interpreted as having the same shape as Riemann's "spherical space" has been noticed independently by several authors. The earliest one I know of is the mathematician Andreas Speiser in *Klassiche Stücke der Mathematik,* Orell Füssli, Zürich, 1925, pp. 53–59. A more detailed analysis is given by the physicist Mark Peterson in "Dante and the 3-sphere," *American Journal of Physics,* Vol. 47 (1979) pp. 1031–35. For more on this subject, and on a finite vs. an infinite universe, see Rudy Rucker's *Infinity and the Mind,* Bantam, New York, 1983, pp. 16–23.

p. 90 "a *hypersphere*": Mathematicians use expressions such as "hyperplane," "hypercube," and "hypersphere" to refer to higher-dimensional versions of a plane, cube, or sphere. Most often, there is a further assumption that the object in question lies in a space of one higher dimension. Four-dimensional euclidean space is most easily described by means of coordinates; just as points in the plane can be characterized by a pair of coordinates (x, y) and points in space by a triple (x, y, z), so points in four-space are given by a quadruple of real numbers, say $(-1, 2, \sqrt{3}, \pi)$, where the separate numbers: -1, 2, $\sqrt{3}$, and π are called the "coordinates" of the point. The equation $x^2 + y^2 = r^2$ defines a circle of radius r in the plane; $x^2 + y^2 + z^2 = r^2$ defines a sphere of radius r in the space; and $x^2 + y^2 + z^2 + w^2 = r^2$ defines a "hypersphere" of "radius" r in "euclidean 4-space." Many properties of hyperspheres can be derived most easily from this equation, but one doesn't *need* the model using four-space in order to study the hypersphere.

p. 91 "Max Born": The quoted excerpt is from the proceedings of the *Jubilee of Relativity Theory (Fünfzig Jahre Relativitätstheorie)* published as Supplement IV to *Helvetica Physica Acta,* Birkhäuser Verlag, Basel, 1956, p. 254.

Chapter VI

p. 93 "grouping of years into decades": See the essay "Decades" by Nancy K. Miller in *Changing Subjects,* edited by Gayle Greene and Coppélia Kahn, Routledge, New York, 1993, pp. 31–47.

p. 94 "just because it is a multiple of ten": An unlikely congruence of politics, geodesy, and the decimal system provides one of the stranger footnotes to the story of how we arrived at our modern units of measurements. A *meter* was defined to be one ten-millionth of the distance from the equator to the North Pole. In other words, the circumference of the earth, measured over the poles, was *defined* to be forty million meters. In 1791, in what may have been the first example of government-financed "big science," a geodetic survey was launched, to determine the exact length of the meridian through Paris from the English Channel to the Mediterranean. That would in turn be used, via astronomical measurements, to determine the length of one degree of latitude, in much the same manner employed by al-Khwarizmi almost a thousand years earlier. Since there are ninety degrees of latitude from the equator to the pole, the resulting figure would determine the length of the meter—one ten-millionth of the total distance.

A fuller account of the reasoning and the machinations that led to the units of measurements now adopted around the world can be found in "The Politics of the Meter Stick" by John Heilbron, *American Journal of Physics,* Vol. 57 (1989), pp. 988–92.

p. 94 "in the year 1800": Whether the nineteenth century begins on January 1, 1800, or January 1, 1801, is—believe it or not—the subject of continuing debate. For recent analyses, see the book *Century's End* by Hillel Schwartz, Doubleday, 1990, and the article by Stephen Jay Gould, "Dousing Diminutive Dennis' Debate," in *Natural History,* April 1994, pp. 4–12.

p. 97 "Maxwell's equations": To paraphrase a popular T-shirt:

And God said

$$\nabla \times H = \epsilon \frac{\partial E}{\partial t} \qquad \nabla \times E = -\mu \frac{\partial H}{\partial t}$$

$$\nabla \cdot H = 0 \qquad \nabla \cdot E = 0$$

. . . and then there was light

There are many different forms of Maxwell's equations. In this particular form, the remarkable and unexpected symmetry between the electric field E and the magnetic field H becomes apparent. By combining all four equations, one can show that each field E and H separately satisfies the "wave equation" governing the motion of a wave traveling in space.

p. 98 "I hold to be great guns": The importance of Maxwell's discovery only grew with the passing of time. In 1938, Einstein and Infeld wrote in their book *The Evolution of Physics:* "The theoretical discovery of an electromagnetic wave spreading with the speed of light is one of the greatest achievements in the history of science." For more on Maxwell, see L. Campbell and W. Garnett, *The Life of James Clerk Maxwell,* London, 1882, and for a brief biography with an excellent account of his scientific contributions: C. W. F. Everitt, *James Clerk Maxwell, Physicist and Natural Philosopher,* Scribners, New York, 1975.

p. 99 "a new era for astronomy": It was Karl Jansky who first identified radio waves from space while searching for the source of static in radiotelephone transmissions. During the early thirties, Jansky tried to further pinpoint the origin of his interference, and he concluded that it was largely from the Milky Way. However, neither he nor anyone else followed up these initial discoveries until Reber built—with his own hands and his own money—the first steerable radio dish, thirty-one feet in diameter, and made a systematic radio-frequency survey of the sky.

p. 99 "Wheaton, Illinois": Wheaton is a suburb of Chicago, where Reber worked as a radio engineer. The full reference to his article "Cosmic Static" is: *Astrophysical Journal,* Vol. 100, 1944, pp. 279–87. The subsequent history of radio astronomy is described in *The Invisible Universe Revealed: The Story of Radio Astronomy* by Gerrit L. Verschur, Springer, New York, 1987. An excellent survey of all the tools of modern astronomy and what they tell us can be found in *The Astronomer's Universe: Stars, Galaxies and Cosmos,* by Herbert Friedman, Ballantine Books, New York, 1990.

Chapter VII

p. 104 "attributed to Edwin Hubble": For more on Hubble's life and contributions, see "Edwin Hubble and the Expanding Universe" by Donald E. Osterbrock, Joel A. Gwinn, and Ronald S. Brashear, *Scientific American,* July 1993, pp. 84–89.

p. 104 "a wonderful interplay between theory and observation": One of the best discussions that I have seen of the steps toward establishing the expansion of the universe and the formulation of Hubble's Law is the recent book by P. J. E. Peebles, *Principles of Physical Cosmology,* Princeton University Press, Princeton, 1993, pp. 77–82. On pp. 82–93 there is an equally thorough discussion of the observational evidence in favor of Hubble's Law. Another excellent exposition is Chapter 10 of Edward R. Harrison's *Cosmology: The Science of the Universe,* Cambridge University Press, 1981. For lots of historical detail, especially on the decades-long attempt to determine whether the "spiral nebulae" were within our own galaxy or were external "island universes" of their own, see *The Expanding Universe: Astronomy's "Great Debate" 1900–1931* by Robert W. Smith, Cambridge University Press, 1982. Finally, for a beautifully clear and lively presentation of the subject by one of the participants in the early debates, see Sir Arthur Eddington, *The Expanding Universe,* Cambridge University Press, 1933 (reissued 1987).

p. 104 "first hints that the universe was expanding": The idea of the expanding universe was in itself an amalgam of theory and observation. The observational component was that the light from distant galaxies was shifted toward the red end of the spectrum, and that the more distant the galaxy the greater the shift. The theoretical part was interpreting the redshift as a kind of Doppler effect to indicate that those galaxies were moving away from us at speeds that increased with the distance to the galaxy. After the theoretical prediction of a redshift by de Sitter in 1917, the topic was discussed actively by leading astronomers such as H. N. Russell and Harlow Shapley at least as early as 1920 (see Smith, pp. 175–76). Papers attempting to establish a relation between redshift and distance based on available observations were published by Ludvik Silberstein and Knut Lundmark during the period 1923–25 (see Smith, pp. 175–78). Taken together with other efforts throughout the 1920s, both observational and theoretical, to try to establish first the reality and second the meaning of de Sitter's 1917 prediction of a redshift-distance relation, they constitute a body of work that makes all the more mysterious the myth of Hubble's sudden discovery of this relation in 1929.

p. 104 "Albert Einstein and . . . Willem de Sitter": Einstein and de Sitter's papers are reprinted, along with a number of other seminal papers on cosmology, in the book *Cosmological Constants,* edited by Jeremy Bernstein and Gerald Feinberg, Columbia University Press, New York, 1986.

p. 105 "Alexander Friedmann": Friedmann was both an exceptionally broad scientist and a colorful character who died in his thirties from typhoid fever contracted in an out-of-the-way area where he landed after a risky balloon ascent. A full-length biography has recently been translated into English: *Alexander A. Friedmann: The Man Who Made the Universe Expand,* by Edward A. Tropp, Victor Ya. Frenkel, and Artur D. Chernin, translated by Alexander Dron and Michael Burov, Cambridge University Press, 1993. Friedmann gained particular notoriety when Einstein first stated in print that he thought Friedmann's calculations were wrong and then later was forced to retract the statement and admit in print that Friedmann was right after all.

p. 105 "Hermann Weyl": Weyl became, along with Einstein, one of the mainstays of the prestigious Institute for Advanced Study in Princeton after its founding.

p. 106 "Harlow Shapley who first proclaimed": See Chapter 2 of Smith on Shapley's model of the galaxy.

p. 107 "Georges Lemaître . . . and Howard Robertson": For more on Lemaître, Robertson, and their contributions, see Harrison's *Cosmology.*

p. 108 "the retroverse": The more usual terms for the retroverse are the "backward light cone" or "null cone." These terms are also used, however, in a somewhat different fashion. (The retroverse is a part of space-time—the actual universe. The "light cone" or "null cone" is often considered to lie not in space-time itself, but in what is called the "tangent space" to space-time.)

p. 110 "Current best estimates": There is some disagreement about the precise value of Hubble's constant. However, the largest value among those generally proposed is only about twice the smallest, so the range of possible values is fairly limited. In any case, the uncertainty in the precise value has no bearing on the correctness of the general picture described here. For a recent discussion of some of these issues, see John P. Huchra, "The Hubble Constant," *Science,* Vol. 256 (17 April 1992), pp. 321–25.

p. 115 "knowing what the precise shape is": There are both theoretical and practical problems with establishing the shape of the slice of the universe that we see looking out in a circle around the horizon. On the practical level, we have no way to measure distances between any two points in our field of view. We can directly measure the angle between our lines of sight to two points, and we can estimate the distance from us to each of them. Knowing those two distances and the angle between them is like knowing the "side-angle-side" of a triangle in plane geometry. The length of the third side of the triangle would be the desired distance between the two points that we are observing. *If* euclidean geometry were applicable, then we could easily determine the unknown distance by standard trigonometric formulas (the "cosine law"). Similarly, if hyperbolic or elliptic geometry were applicable, known formulas would provide the answer in each case. The problem is that we do not know the geometry in advance, and what we are trying to do is to deduce as much as we can about that geometry by making whatever measurements we can, just as Gauss was doing when he proposed determining as much as possible about the geometry of a surface by making measurements along the surface.

On the theoretical side, if we accept Einstein's general theory of relativity, then even the concepts of "distance" and "time" are not well-defined, and only a certain combination of them has invariant meaning. That combination is equal to zero along what is called the "null cone" or the "backward light cone," which corresponds exactly to what we have called the "retroverse": all those points "visible" to us at a given instant of time via light or other forms of electromagnetic radiation. So measuring lengths of curves on the retroverse does not seem possible in the context of general relativity.

However, the situation is not quite as bleak as it initially appears. The proscriptions on what can and cannot be measured as decreed by the theory of relativity refer to measurements made from *within* the system, again in the manner of Gauss surveying a surface, or Riemann describing the geometry of space. In the case of the entire universe, it would seem self-evident that the only measurements possible are those within the universe. But it turns out that both in theory and in practice, there is a way to refer measurements to an "event" that lies outside our universe. That event is the big bang. The big bang is often described as a singularity in space-time, which tends to cloak the fact that it is not itself a point of space-time. But it *is* a point to which all points of space-time may be referred.

Twentieth-century cosmology, in its attempts to describe the shape of the universe, has generally focused on constructing models of the universe in the form of four-dimensional space-times where space evolves in time, starting at the big bang. In these models, "time" is well-defined: time since the big bang, where again, the big bang is not a point of space-time, but a reference point outside providing the zero point of time, just as "absolute zero" (zero on the Kelvin scale) does not correspond to an attainable temperature in the real world, but a point off the scale to which all other temperatures are referred. Although it was not obvious when the original theoretical models of the universe were constructed, there is even a direct experimental way for us to measure time as understood in the models: through the temperature of the cosmic microwave background radiation. As the universe expands, the background radiation cools down, and in any given model the temperature of that radiation can be related to the time since the big bang. Since we can directly measure the temperature of the radiation, we can derive a measurement of time.

A related benefit of the microwave background radiation, available in our real universe but not as a part of general relativity, is that it permits us to describe the "present universe"—the space component of space-time corresponding to a moment of earth-time: "now." The "clock" we use to measure time at any point of the universe is the temperature of the background radiation, and the "present universe" consists of all points of space-time for which that temperature is the same as we measure it to be now on earth.

Hubble's Law is often described as a relation between recession velocity and distance of galaxies—specifically, that the ratio of velocity to distance is constant. However, neither "velocity" nor "distance" is determined by direct observation. What can be measured is the redshift in light from each galaxy, and theory is then used to convert the redshift to velocity. As for "distance," most modern treatments of Hubble's Law interpret distance between galaxies as "present distance," meaning how far apart they are in the present universe. Although the present universe can be clearly defined in the way we have described, using background microwave radiation temperature, there does not seem to be any comparable way of determining present distance. Each model of the universe as space evolving with time includes a notion of present distance in space, and in each model Hubble's Law holds with both "distance" and "velocity" referred to present distance in that model.

One of the standard models is called the *Einstein–de Sitter model* in which space is euclidean and expands at a well-determined rate. In this model we can describe the precise shape of the retroverse. In our map of the retroverse, we drew a circle corresponding to all galaxies at a given distance from us, lying along the horizon. Since we are seeing those galaxies at a specific time in the past, that circle lies in a euclidean space (in the Einstein–de Sitter model), and its length is $2\pi r$, where r is the distance *in that euclidean space* between those galaxies at the time the light we now see was emitted, and the position of the earth *at that time.* Carrying out the calculations, we arrive at a surface whose shape is very much like an upside-down top. It is the surface obtained by rotating the curve $x = 3[z^{2/3} - z]$, $0 \leq z \leq 1$, about the z-axis. If T is the time from the big bang to the present, then $t = Tz$ represents the time from the big bang that light reaching us now left a certain galaxy, and $r = Tx$ is the distance between us and the galaxy at that time.

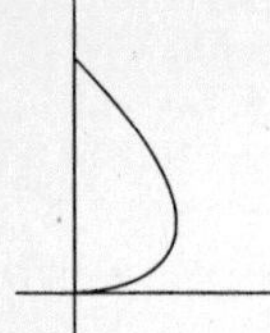

THE CURVE $x = 3(y^{2/3} - y)$, $0 \leq y \leq 1$.

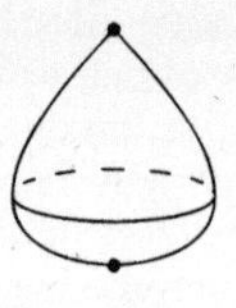

THE SURFACE OBTAINED BY ROTATING THE CURVE ABOUT THE Y-AXIS.

The shape of the retroverse looking out in all directions along the horizon in an Einstein–de Sitter universe.

p. 119 "some fairly convincing physical evidence": See, for example, Chapter 4: "Evidence for the Big Bang," in the book by Joseph Silk, *The Big Bang,* W. H. Freeman, New York, 1989. Two of the most persuasive bits of physical evidence are the observationally confirmed predictions of the cosmic microwave background radiation and the distribution of helium and deuterium in the universe. Both predictions are based on the physics of the early universe in a big bang scenario, and nobody has found a convincing alternative argument for the observations.

p. 122 "a choice familiar to world-map watchers": The map used is called the "Hammer projection" or "Hammer-Aitoff projection." It represents the world inside an ellipse. A Hammer map of the earth is shown in Chapter 2.

p. 123 "The LIGO project": For a fascinating account of the LIGO project and the associated physics, see the book, *Black Holes & Time Warps: Einstein's Outrageous Legacy,* by Kip S. Thorne, W. W. Norton, New York, 1994.

Chapter VIII

p. 125 "Albert Einstein": The event described here is based on the personal recollections of the author, who attended the conference in Princeton as a graduate student working on the theory of Riemann surfaces.

p. 125 "the 'Riemann surface'": Riemann surfaces were the first examples of "abstract surfaces," discussed in Chapter 9. The notion has evolved over the years, but as originally defined by Riemann, they are formed by taking a number of copies of the standard plane—they can be pictured as a set of parallel planes in space, or more concretely, as a number of sheets of paper piled on top of each other—and, after cutting them along certain lines, "decreeing" that they be attached to each other in ways that are not possible in reality, but that are perfectly feasible abstractly. (See the examples of abstract surfaces described in Chapter 9.) The simplest example of a Riemann surface consists of two "sheets"—copies of the plane—each cut along the same ray starting at a point and extending to infinity, then attached to each other "crosswise," the two edges of the cut in the top sheet attached to the opposite edges of the cut in the bottom sheet. That Riemann surface turns out to be very useful in studying complex numbers—combinations of real and imaginary numbers—and their square roots.

p. 127 "Einstein's wry remarks and aphorisms": Ronald W. Clark's *Einstein: The Life and Times,* World Publishing, New York, 1971, reports on many of Einstein's sayings, as well as his scientific and personal life. The often-cited "I cannot believe that God plays dice with the universe" does not appear to be a direct quotation (see Clark, p. 340). "Subtle is the Lord . . ." is an attempt to render into English Einstein's actual words *("Raffiniert ist der Herrgott, aber boshaft ist er nicht").* For views on religion and science complementary to those of Einstein, see the allocution of Pope John Paul II in *Theory and Observational Limits in Cosmology,* Specola Vaticana, Vatican City, 1987, pp. 17–19.

p. 128 "Einstein's 'special theory of relativity'": A translation of Einstein's original paper, entitled "On the electrodynamics of moving bodies," appears in the collection *The Principle of Relativity,* Dover, New York, 1952. It makes excellent reading, and is probably as clear, or clearer, than many "popular" accounts, for anyone comfortable with the use of elementary mathematics.

p. 129 "a case can be made": The argument that Einstein was the true originator of quantum discontinuity is made very effectively by Thomas Kuhn in his book *Black-Body Theory and the Quantum Discontinuity, 1894–1912,* University of Chicago Press, 1987.

p. 129 "Hermann Minkowski": Minkowski's interpretation of special relativity in terms of a four-dimensional space-time was presented in the form of a lecture in Cologne, Germany, in September 1908. An English translation, entitled "Space and Time," can be found in the collection *The Principle of Relativity,* Dover, New York, 1952.

p. 132 "a convenient abstraction": See the article by Hubert Reeves, "Birth of the myth of the birth of the universe," in *New Windows on the Universe,* Vol. 2, Eds. F. Sanchez and M. Vazquez, Cambridge University Press, Cambridge, 1990, pp. 141–49.

p. 136 "space was a passive background": The main precursor to Einstein in this area was the brilliant young English mathematician, William Kingdon Clifford, whose contributions were cut short when he died at age 33 on March 3, 1879, just eleven days before Einstein was born. Clifford was strongly influenced by Riemann. He arranged for the publication of an English translation of Riemann's basic paper on curved space and related notions, and he propagated Riemann's ideas both at scientific meetings and in public lectures. Chapter 4 of his book, *The Common Sense of the Exact Sciences,* includes a section "On the Bending of Space" which provides a beautiful exposition of the notion of curved space, concluding with the statement, "We might even go so far as to assign to this variation of the curvature of space 'what really happens in that phenomenon which we term the motion of matter'." (The situation is a bit more complicated in that Clifford's book was published posthumously under the stewardship of Karl Pearson who actually wrote all of Chapter 4. However, a footnote to the passage cited above says that "This remarkable *possibility* seems first to have been suggested by Professor Clifford in a paper presented to the Cambridge Philosophical Society in 1870 *(Mathematical Papers,* p. 21).") In any case, although Riemann and Clifford may have glimpsed the possibility of linking the curvature of space to physical phenomena, it remained for Einstein to formulate a precise set of mathematical equations governing that linkage. (We should also mention Karl Schwarzchild, who published a paper in 1900 in which he took quite seriously the curvature of space, and computed quantities such as the parallax of stars in curved space. Later on, Schwarzschild found the first solution to Einstein's equations of general relativity.)

Chapter IX

p. 142 "This phenomenon": Einstein also addressed the question in a lecture on January 27, 1921, at the Prussian Academy of Science: "At this point an enigma presents itself which in all ages has agitated inquiring minds. How can it be that mathematics, being after all a product of human thought which is independent of experience, is so admirably appropriate to the objects of reality?"

p. 144 " 'buckyballs' ": An excellent article on the mathematics of buckyballs is Fan Chung and Shlomo Sternberg's "Mathematics and the Buckyball," *American Scientist*, Vol. 81 (January–February, 1993), pp. 56–71.

p. 153 "Clifford torus": The equations $x = \cos u \cos v$, $y = \cos u \sin v$, $z = \sin u \cos v$, $w = \sin u \sin v$, $0 \leq u \leq 2\pi$, $0 \leq v \leq 2\pi$, define a surface lying in the hypersphere $x^2 + y^2 + z^2 + w^2 = 1$. That surface is the Clifford torus.

p. 156 "made greater sense as an abstract model": The issue of a finite vs. an infinite universe has been debated down through the ages. The main lines of the debate from ancient times through the period of Newton and Leibniz are well laid out in the book *From the Closed World to the Infinite Universe* by Alexandre Koyré, Johns Hopkins Press, Baltimore, 1957. Twentieth-century cosmology has altered the terms of the debate, but by no means ended it. See, for example, J. D. North, *The Measure of the Universe,* Clarendon Press, Oxford, 1965, especially Chapter 17. A more recent attempt to analyze the consequences of a physically infinite universe is "Life in the Infinite Universe" by G. F. R. Ellis and G. B. Brundrit, *Quarterly Journal of the Royal Astronomical Society,* 20 (1979), pp. 37–41. It would seem that the origins of the universe in the big bang would lend strong support to the notion that what emerges from the big bang would be finite in size and content. However, somewhat surprisingly, an informal poll of leading cosmologists reveals that a sizeable majority finds no problem with accepting a physically infinite universe, with an infinite number of stars and galaxies distributed through an infinite expanse of space. It could be argued that the debate is philosophical or metaphysical rather than scientific, since the *observable* universe *is* finite, and it is not clear if there will ever be an experimental way of determining whether what lies beyond our limits of observation is finite or infinite in extent. One of the most thoughtful commentators on these issues is G. F. R. Ellis; see his article "Major Themes in the Relation between Philosophy and Cosmology" in *Memorie della Societa Astronomica Italiano* 62 (1991), pp. 553–605. An interesting, if somewhat more fanciful, discussion of many of these matters can be found in the book *Infinity and the Mind* by Rudy Rucker, Bantam, New York, 1983.

p. 159 "another equally valid option": The alternatives to a spatially infinite universe in the form of flat or hyperbolic manifolds are often referred to in cosmology as "small universes" or "periodic universes." Some properties of such models, and how we might detect it if our universe were actually shaped like a flat torus or a hyperbolic manifold, can be found in the papers "An Introduction to Small Universe Models" by Charles C. Dyer, "Observational Properties of Small Universes" by G. F. R. Ellis, and "Observational Constraints on 'Small Universes' " by R. B. Partridge, all in the volume *Theory and Observational Limits in Cosmology,* Proceedings of the Vatican Observatory Conference held in Castel Gandolfo, edited by W. R. Stoeger, S. J., Specola Vaticana, Vatican City, 1987, pp. 467–88.

p. 159 "one of the many finite-sized hyperbolic manifolds": For a discussion of hyperbolic and other manifolds in the context of the possible shape of the universe, see "The Mathematics of Three-dimensional Manifolds" by William P. Thurston and Jeffrey R. Weeks, *Scientific American,* July 1984, pp. 108–20.

p. 160 "Benoit Mandelbrot": Mandelbrot's book was originally written in French; an English translation appeared in 1977. An updated version is *The Fractal Geometry of Nature* by Benoit B. Mandelbrot, W. H. Freeman, New York, 1983.

p. 166 "an exact value for the dimension": If a figure has dimension d, then scaling the figure up by a factor of 3 will increase the measure of the figure by a factor 3^d. For a curve, with $d = 1$, this factor is 3. For a surface, with dimension $d = 2$, the factor is 3^2 or 9. For a 3-dimensional figure the factor is 3^3, or 27. In the case of the snowflake curve, the "size" or measure increases by a factor of 4. Hence $3^d = 4$, which means $d \log 3 = \log 4$, or $d = (\log 4)/(\log 3)$, which is the exact dimension of the snowflake curve.

p. 166 "Mandelbrot later interpreted": See *The Fractal Geometry of Nature,* Section 9: "Fractal View of Galaxy Clusters."

p. 167 "attempts to reconcile a fractal structure": For a detailed account, taking into consideration all the evidence through the early 1990s, see P. J. E. Peebles, *Principles of Physical Cosmology,* Princeton University Press, Princeton, 1993, pp. 209–24: "Fractal Universe and Large-Scale Departures from Homogeneity."

Postlude

p. 169 "Richard Feynman": The character of Richard Feynman is revealed in his autobiographical memoir, *Surely You're Joking, Mr. Feynman,* Bantam, New York, 1986, and in the recent biography, *Genius,* by James Gleick, Vintage, New York, 1993.

p. 169 "*The Character of Physical Law*": M.I.T. Press, Cambridge, Massachusetts, 1967, p. 39 and p. 58. Similar views have been expressed by another leading twentieth-century physicist, Paul Dirac, who added, "A physical law must possess mathematical beauty." See the article "P. A. M. Dirac and the Beauty of Physics" in *Scientific American,* May 1993, pp. 104–9.

Index